D0850201

PHalarope books are designed specifically for the amateur naturalist. These volumes represent excellence in natural history publishing. Each book in the PHalarope series is based on a nature course or program at the college or adult education level or is sponsored by a museum or nature center. Each PHalarope book reflects the author's teaching ability as well as writing ability. Among the books:

The Amateur Naturalist's Diary
Vinson Brown

Nature with Children of All Ages: Activities and Adventures for Exploring, Learning, and Enjoying the World Around Us
Edith A. Sisson, the Massachusetts Audubon Society

The Plant Observer's Guidebook: A Field Botany Manual for the Amateur Naturalist
Charles E. Roth, Massachusetts Audubon Society

The Wildlife Observer's Guidebook
Charles E. Roth, Massachusetts Audubon Society

Eric V. Gravé is a naturalist and nature photographer specializing in photography through the microscope. He has served as curator and as a member of the board of directors of the New York Microscopical Society and was on the staff of the College of Physicians and Surgeons, Columbia University, for twenty-five years.

Discover the Invisible

A NATURALIST'S GUIDE TO USING THE MICROSCOPE

Eric V. Gravé

Prentice-Hall, Inc.,
Englewood Cliffs, New Jersey 07632

Library of Congress Cataloging in Publication Data

Gravé, Eric V.
Discover the invisible.

(PHalarope books)
"A Spectrum Book."
Bibliography: p.
Includes index.
1. Microscope and microscopy. I. Title. II. Series.
QH205.2.G73 1984 578 84-6777
ISBN 0-13-215344-0
ISBN 0-13-215336-X (pbk.)

COVER: A living Paramecium caudatum *photographed with an interference microscope after Jamin-Lebedeff. It reveals numerous cell organelles otherwise not visible unless the animalcule is killed and chemically stained. For details see Plate 1 and page 50. The background of the cover shows the cells of petrified wood in cross section photographed with an interference microscope after Nomarski. For details see Figure 10–5 page 175.*

A Spectrum Book. Printed in the United States of America.

10 9 8 7 6 5 4 3 2 1

ISBN 0-13-215344-0

ISBN 0-13-215336-X {PBK.}

Editorial/production supervision by Marlys Lehmann
Book design by Maria Carella
Page layout by Debra Watson
Cover design by Hal Siegel
Manufacturing buyer: Frank Grieco

This book is available at a special discount when ordered in bulk quantities. Contact Prentice-Hall, Inc., General Publishing Division, Special Sales, Englewood Cliffs, N.J. 07632.

Prentice-Hall International, Inc., *London*
Prentice-Hall of Australia Pty. Limited, *Sydney*
Prentice-Hall Canada Inc., *Toronto*
Prentice-Hall of India Private Limited, *New Delhi*
Prentice-Hall of Japan, Inc., *Tokyo*
Prentice-Hall of Southeast Asia Pte. Ltd., *Singapore*
Whitehall Books Limited, *Wellington, New Zealand*
Editora Prentice-Hall do Brasil Ltda., *Rio de Janeiro*

To the memory of my wife,
Lily,
who contributed to this book
in many ways

Contents

Before We Start, xiii

PART ONE

1
The Early Micronauts, 3

2
Your Microscope, 11

3
Special Methods of Illumination, 33
Brightfield Illumination, 33
Darkfield Illumination, 33
Oblique Illumination, 39
Rheinberg Illumination, 41
Light Polarization, 43
Incident Illumination, 46
Modulation Contrast, 47
Phase Contrast, 48
Interference Microscopes, 49
Fluorescence, 51

4
Your Specimens, *54*

5
Photomicrography, *68*

PART TWO

6
"I Eat—Therefore I Am," *85*
Chaos chaos Catching Paramecium, 87
Paramecium in the Robber's Den, 90
Blepharisma, the Cannibal, 93
About a Unique Animalcule, 94
Lacrymaria, the Swananimalcule, 98
Hydra Catching Waterflea, 100
A Plant That Captures Animals Under Water, 100
The Venus's-Flytrap, 106

7
Of Children and Children's Children, *111*
An Animalcule Divides, 111
Where Quadruplets Are the Rule, 113
Didinium's Survival, 114
Conjugating Paramecia, 119
Cyclops in Production, 123
Planaria—the Amazing Worm, 125
The Vinegar Eel, 129

8
Of War, Peace, and How to Hitch a Ride, *132*
Warfare in the Microworld, 132
Life with an Alga, 136
Life in the Hindguts of Wood-Feeding Insects, 143
Protozoa in the Rumen of Ruminants, 150
Pilobolus, the Sharpshooting Fungus, 152
Some Hitchhiking Plants and Animals, 156

9

Some Aspects of Behavior Revealed By the Microscope, 160

When Stentor Gets Scared, 160
Fish Scales, 163

10

A Trip to the Parks, 169

The Hot Spring Algae of Yellowstone, 169
The Petrified Forest of Arizona, 172

Resources, 176
References, 179
Glossary, 185
Notes, 193
Index, 197

Before We Start

This book is a guide to a seemingly invisible universe—the microworld. It is mostly unknown to the majority of us because of the limitations of our vision. Our eyes are marvelous organs, yet they do not permit us to read a road sign if it is too far away. The same is true in the other direction; we are also limited in our sight if the object of our interest is brought too close to our eyes. For instance, if you lift this book to your eyes to inspect the appearance of a period at close range, you would not see it bigger or more distinctly. To the contrary, the print would become fuzzy.

To point out ways to overcome our vision handicap is the purpose of this book. The necessary aid to our vision is the microscope; the task ahead is to explain its use and to show what can be seen with it. This is an instructive and entertaining effort. It widens the horizon of our knowledge of the hidden part of nature. In observing microscopic plant life, we can discover unexpected beauty, and in the animal world, amazing happenings—the different ways the little creatures in the pond feed, reproduce, fight, or live together for their mutual benefit. In the material world revealed by the microscope, we are bound to discover an endless variety of artistic patterns and colors.

The microscope as an instrument is not older than 350 years. Before that time, none of our ancestors knew anything about the invisible world. During the seventeenth and eighteenth centuries, when word got around about the microscopes, the instrument was accessible only to the wealthy. Today we are fortunate in that microscopes are within the reach of those with even modest savings accounts. Because of the diversity and range the instrument has to offer, it is of interest to all age groups: fledgling high school students who may be stimulated to choose microscopy for their careers; adult nature lovers; senior citizens who need something to occupy their minds; photographers in search of new motifs for their cameras.

In this book, microscopists will sometimes be called *micronauts* because searching for the invisible is like orbiting the microcosm much like the astronauts orbit the greater cosm. But what benefits can readers expect by becoming micronauts?

1. Micronauts widen their outlook on life.
2. They can entertain friends by showing them the interesting things discovered with the help of the "looking glass."
3. They can engage in an active and creative endeavor because some of the specimens to be observed have to be prepared, crystals must be grown, and one-celled animalcules cultured.
4. The microscope can offer opportunities to do research, if only of modest scope. As the years go by, the micronaüt becomes more familiar with what is known and what is new. In this respect, I myself was once fortunate to discover, in an insect, microscopic animals never described before.
5. Photomicrography, the art of photographing through the microscope, may provide the best chances for creativity and success. A nature photographer records a given image as it is. Sometimes nature helps by providing a dramatic sunset, a spectacular mountain range, a beautiful flower, butterfly, or other striking jewel of the visible world. In such cases, it is nature that produces the appealing image; not so in photomicrography. For instance, crystals that you have grown yourself can, though colorless themselves, be photographed in striking colors using simple techniques, which are described in later chapters.

The first chapter provides a brief early history of the microscope. The following chapters offer instructions on how to set up a microscope for regular observation, special methods of illumination, instructions on how to handle easily available objects from our environment, and, finally, directions on how to photograph through the microscope.

The rest of the book is devoted to "discoveries." Adventuring always holds an element of surprise. When you set out to roam about in the microworld, you enter a new dimension. This dimension is the wonderful aspect of the *infiniment petit,* as the French philosopher Blaise Pascal (1623–1662) called the infinitely small: that there are thousands of new experiences in store, some aesthetic, others instructive or puzzling, yet always fascinating.

There is no intention, nor is it possible, to give a comprehensive or systematic account of the microcosm. Yet many observations reported in these pages are not even mentioned in textbooks on the high school or college level. There is, for example, a description of the life story of a rare worm that can be found only on the coast of Brittany, France, and nowhere else in the world. To observe and photograph it required a trip to France, loaded with all the microscopical and photographic equipment necessary for this expedition. Another example are the fauna in the rumen of cattle, known to specialists, but rarely mentioned in publications available to the general public. One of the most remarkable plants existing on our globe, a tiny mushroom that grows on horse manure, is often overlooked even by biologists.

This book presents little-known elements of the microworld. In this respect, I have greatly benefited from the remarkable technological progress made during the past few decades. Many illustrations of living organisms would not have been possible without the invention of the electronic flash. Furthermore, the development of phase contrast, a special mode of illumination of microscopic objects, made it possible to take pictures that provide more information than those taken with conventional optics.

I hope that you will profit from and enjoy these "discoveries" and be spurred on to some of your own.

Acknowledgments

Any published book is a product of efforts in which the author's editor plays a vital role. In this sense, I am deeply indebted to my editor, Mary E. Kennan, whose valuable suggestions are incorporated in this book. Only an author knows an editor's contribution since her or his work remains anonymous, invisible to the reader of the final opus, and no microscope, alas, can make it visible. It was Mary Kennan who suggested I make a guide out of a manuscript that previously contained instructions only as brief appendices, thus making it a useful book.

I am also much obliged to Marlys Lehmann, my production editor, whose able, patient, and conscientious work has greatly helped to shape this book.

Many friends have shared their knowledge with me. I think especially of Theodore G. Rochow, professor emeritus at the University of North Carolina, author of a book on microscopy, who read the technical part of the manuscript; Professor Margaret R. Murray, pioneer in the field of tissue culturing; Dickson D. Despommier, professor of tropical medicine, Columbia University; Margaret G. Cubberly, Fellow and past president of the New York Microscopical Society and a superb histologist; and Fritz Goro, the eminent scientific photographer.

I am much obliged to various institutions, above all to the New York Microscopical Society and the Biological Photographic Association. Their many programs—lectures, workshops, field trips, gossip meetings, and conventions offered invaluable stimulants and the possibility to exchange knowledge and experiences that as a whole were sources of learning not otherwise available. I also would like to express my appreciation to the College of Physicians and Surgeons, Columbia University, where I enjoyed the priviledge of laboratory facilities for many years.

My thanks go also to Zena Toran, who provided needed artwork, and Jeanne D. Cole, who assisted in preparing the manuscript.

Finally, I wish to thank the editors of those periodicals who permitted me to include in this book articles that were previously published in their magazines' pages: Alan Ternes of *Natural History*, D. R. Edwards of *Microscopy* (London), and Dr. Dieter Krauter of *Mikrokosmos* (Stuttgart, West Germany).

Great and gorgeous as is the display of divine power and wisdom in the things that are seen of all, it may safely be affirmed that a far more extensive prospect of these glories lay unheeded and unknown, til the optician's art revealed it. Like the work of some mighty genie of oriental fable, the brassy tube is the key which unlocks the world of wonder and beauty before invisible, which one who has once gazed upon it can never forget, and never cease to admire.

From Philip Henry Gosse, F.R.S.,
Evenings at the Microscope,
London, 1859

part
one

1
The Early Micronauts

Only about 400 years have passed since the first primitive microscopes were developed. It has now been established that the ancient Egyptians, Babylonians, Greeks, and Romans had no means of viewing objects at even the lowest magnifications. The very first mention of the phenomenon of magnification appeared in the writings of the Roman statesman and philosopher, Seneca (4 B.C.–A.D. 65). He stated that "letters though small and indistinct are seen enlarged and more distinct through a globe filled with water."[1] However, he concluded in error that the enlargement was due to the water, whereas actually the water-filled vessel acts as a lens.

After Seneca, progress was slow. A thousand years had to pass before the next small step was taken. Diligent historians, digging in manuscripts of the Dark Ages, uncovered another communication on the magnifying properties of a lens in the writings of Abu Ali al-Hazan ibn Alhasan (962–1038). Alhasan was a physicist and one of the first to investigate the human eye. He found that a piece of glass, curved on one side and flattened on the other, had a magnifying effect if put over an object with the curved side toward the eye. This investigation led to the discovery of the optical properties of a planoconvex lens.

At the end of the thirteenth century, time was ripe for the first spectacles, and here the honor goes to the Franciscan monk, Roger Bacon (1214–1294), of Ilchester, England. In his *Opus Majus*, a book published in 1276, Bacon showed that crystal lenses could be made "instruments useful to old men and those whose sight is weakened, who in such a way will be able to see the letters sufficiently enlarged, however small they are."[2]

Bacon's fate was a tragic one. The church was in conflict with his many original ideas. He spent years in prison and was later confined to a convent of his order, where he died at the age of 80. His writings on natural science had to be hidden to prevent their destruction and were rediscovered only in 1733. Fortunately, there was no delay in providing old men with spectacles, because a Florentine, Salvino degli Amati, developed the idea successfully in 1280.

Three hundred years later, in 1590, a Dutch spectacle maker, perhaps Hans Janssen, had the idea of combining two lenses in such a way that the image magnified by the lens nearest to the object would be further magnified by a second lens near the eye. His instrument, the very first microscope, consisted of telescoped drawtubes with the object-lens on one end and the eye-lens on the other. By regulating the distance between these lenses, the image could be sharpened. This truly outstanding invention was the first compound microscope. Its importance can be gauged by the fact that all modern light microscopes are still compound microscopes, which are built according to this basic principle. The Janssen microscope, which had no stand and no mirror, was apparently hand-held and suitable only for viewing opaque objects.

There is some controversy as to whether Janssen was really the first to invent the compound microscope. Another candidate is Galileo (1564–1642) who, by extending the tube of his telescope, modified it for use as a microscope. One of his contemporaries, Giovanni du Pont, Seigneur de Tarde, described his method in 1614:

> *The tube of the telescope for looking at the stars was not more than two feet in length but to see objects that are very near . . . the tubes must be two or three arms' length. With this tube I have seen flies which look as big as lambs and have learned that they are covered with hair and have very pointed nails by means of which they keep themselves up and walk on glass, although hanging feet upwards by inserting the points of their nails in the pores of the glass.*[3]

Needless to say, Galileo's huge microscope was much too clumsy to be developed further.

By the middle of the seventeenth century, scientists had used the compound microscope to scan many areas of the hidden world. Two books that summarized some of the new discoveries appeared. One book written by the French scientist Pierre Borel (1620–1689), the private physician of Louis XIV, was entitled *Centuria Observationum Microscopicarum.* It was written in Latin, and published in The Hague in 1655. The other book was written in English by Robert Hooke (1635–1703) and came out in 1665. It was entitled *Micrographia, or Some Physiological Descriptions of Minute Bodies Made by Magnifying Glasses with Observations and Inquiries Thereupon.* Now considered a classic, it became available a few years ago in a paperback edition by Dover Publications, New York.

Robert Hooke was an "experimental philosopher," as *Webster's New Collegiate Dictionary* describes him, whose scientific interests were very broad. Among physicists he is remembered as the originator of Hooke's Law, which states that the extension of a spring is proportional to the force producing it. He invented a spirit level and a barometer, not to mention his many suggestions for the optical and mechanical development of the microscope. Originally, Hooke was destined to become a portrait painter, but then he went to Westminster School and later to Oxford University, where he became an instructor. In 1663, soon after the Royal Society of London was established, he became its Curator of Experiments. The Society held weekly meetings, and Hooke was charged to "bring in at every meeting one microscopical observation at least." By 1665, he had provided so many observations that they filled an impressive volume, his *Micrographia.* His early training as a painter came in handy because he could make the illustrations himself, and they are magnificent. He described and pictured insects and their parts, the point of a needle, the edge of a razor, "insects" in rainwater, snow crystals, the texture of cork. He described cork as being made up of "little boxes or cells," which reminded him of a honeycomb. Ever since, the components of any animal or plant tissue have been called "cells."

Hooke's original microscope (see Figure 1-1) was built by Christopher Cock (1630–1696). It was suitable only for rather low power, as the observations presented in his book indicate. He knew that the magnifying power of a bi-convex lens depends on the curvature of its surface. The more globe-like its shape, the higher the magnification it can provide. A perfect globe of glass, rock crystal, or quartz has the highest "power"; and the smaller the globe, the higher the magnification. Hooke made one such microscope himself and even gave instructions on how to make a very small lens, the

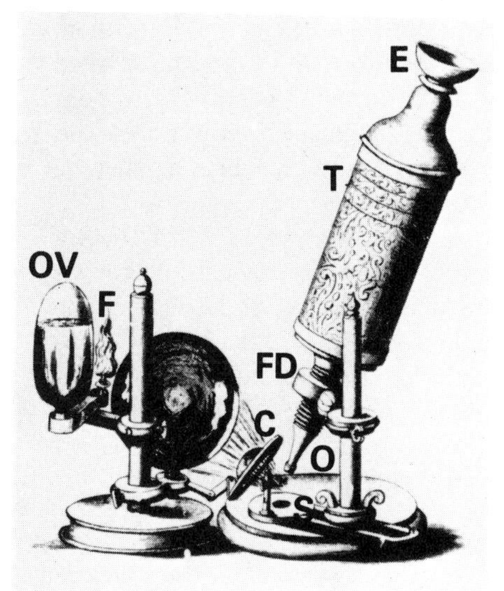

FIGURE 1-1. Robert Hooke's compound microscope. *C* Collecting lens; *E* Eyepiece; *F* flame; *FD* Focusing device; *O* Objective; *OV* Oil vessel; *S* Stage; *T* Tube. *From:* William B. Carpenter, *The Microscope and Its Revelations* 8th edition. (Philadelphia: P. Blakiston's Son & Co., 1901), pg. 110.

single lens or simple microscope he was experimenting with. However, he did not pursue this path because of the inconvenience of using such an instrument. It was necessary to put the lens very close to the eye, as well as close to the object under observation. The strain on the eye from prolonged peering through a tiny lens disturbed him.

Yet there was one man who did not shun the difficulties of using a simple microscope. His persistence, indefatigable zeal, and fascination with the subject enabled him to penetrate much deeper into this new world. He observed things no other scientist of his time had ever observed, and he wrote the first pages of two biological sciences; bacteriology and protozoology, the biology of one-celled plants and animals. His name was Antony van Leeuwenhoek (1632–1723).

Leeuwenhoek was born on October 24, 1632, in Delft, a small town in Holland. His father was a basket maker and his mother was the daughter of a brewer. Descending from a family of tradespeople, he was sent to Amsterdam to learn a trade in a linen-draper's shop. In 1654, he returned to Delft, where he opened his own business as a draper. Six years later, he was made chamberlain to the Sheriffs of Delft, a position that put him in charge of the courtroom "wherein the Chief Judge, the Sheriffs and the Law Officers of this Town do assemble." During the following years, he assumed additional responsibilities in the city administration; having "exercised himself in the art of Geometry," he was permitted "to perform the office and duty of surveyor within the jurisdiction of the Court." Leeuwenhoek was also appointed "wine-gauger." As such, he had "to examine all wines and spirit entering the town and to calibrate the vessels in which they were contained."[4]

Leeuwenhoek probably discharged many of these duties by proxy. This gave him ample time to pursue his chief interest, making tiny lenses. He mounted these in simple microscopes of his own design and with them pioneered the new frontier of the microworld. For many years, he worked entirely on his own. Communications in the seventeenth century were poor and scientists wrote to one another in Latin, a language Leeuwenhoek did not know. One of the most admirable things about this man is that he achieved so much without any scientific training.

The breakthrough came with his first publication in 1673. In 1660, a new scientific society was organized for the "promotion of natural knowledge"—the Royal Society of London. Its first secretary, and editor of the Society's publication, the "Philosophical Transactions," was Henry Oldenburg (1615–1677). Oldenburg was very active in corresponding with scientists in many countries, among them a physician in Delft, Reinier de Graaf (1641–1673), who was an acquaintance of Leeuwenhoek's. It was de Graaf who introduced Leeuwenhoek to the Society with the following letter:

I am writing to tell you that a certain most ingenious person here, named Leeuwenhoek, has devised microscopes which far surpass those which we have hitherto

> *seen. . . . The enclosed letter from him wherein he describes certain things which he has observed more accurately than previous authors, will afford you a sample of his work; and if it please you and you would test the skill of this most diligent man and give him encouragement, then pray send him a letter containing your suggestions, and propose to him more difficult problems of the same kind.*[5]

Leeuwenhoek's first letter was accepted and published in the "Philosophical Transactions" in 1673. His letter contained observations on a mold; on the stinger, mouthparts, and eyes of the bee; and on the louse. Many letters were to follow over the next 50 years. In his eighty-fifth year, he wrote his last published letter, as he said, "with a torpid and trembling hand." In 1680, he was made a Fellow of the Society, and in 1681 he was requested to publish his invention. This he apparently never did.

Leeuwenhoek's microscopes (a drawing of one is shown in Figure 1-2) were not much larger than 1×2 inches. They consisted of two plates, each perforated with a hole, sandwiched together in such a way as to hold a tiny, well-polished bi-convex lens between the perforations. The specimen was placed on a needle that could be moved in all directions in order to bring the specimen into proper focus. He put specimens in fluid on a piece of thin glass, which he then glued to the needle. His design was unique; none of the microscope makers of his time devised anything similar. He never changed its mechanical outline, yet as far as his lenses were concerned, he constantly strived for improvement. They provided a wide range of magnifications, the highest power being 270×(the majority magnified from 40 to 160×). One reason he was so successful was that the design of his device permitted viewing the specimen by holding the specimen against

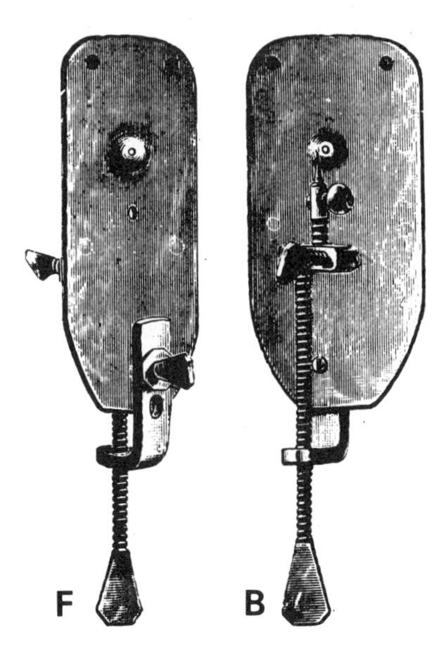

FIGURE 1-2 Leeuwenhoek's simple microscope. *F* Front view; *B* Back view. *From:* William B. Carpenter, *The Microscope and Its Revelations,* 8th edition. (Philadelphia: P. Blakiston's Son & Co. 1901, pg 132.)

the light. The many transparent organisms in the water thus came into view. He was one of the first, maybe the very first, to observe his objects with transmitted light rather than by reflected light, as did his contemporary micronauts.

When Leeuwenhoek died, he left 26 microscopes to the Royal Society, and when his daughter died in 1774, an auction of the remaining instruments was arranged. The auction catalogue listed, according to Clifford Dobell's research, 247 complete microscopes, each with its own lens and many with the object for which they were made. In addition, the catalogue described 172 pieces that consisted only of lenses sandwiched between plates. Three of the complete microscopes had lenses made of rock crystal; one had a lens ground and polished from a sand grain. One hundred and sixty of these microscopes were made of silver, three of solid gold. Of the whole treasure, only nine authentic Leeuwenhoek microscopes have survived, according to Clifford Dobell.

Leeuwenhoek had the habit of examining everything under the sun—another clue to his remarkable success. In the course of exploring this new world, he made an amazing number of new discoveries. By putting some of his own dental plaque under his lens, he made the first discovery of bacteria. Examining his own stool during a bout of diarrhea, he discovered intestinal bacteria and protozoa. He showed that mammals have round red blood corpuscles, while birds, amphibians, and fishes have oval ones. He dissected a shrimp and found a foraminiferan in its stomach; the drawing he ordered his draughtsman to make is so good that it can be identified as a *Polystomella.* Other firsts were *Hydra, Rotifera, Volvox,* and *Vorticella,* to name only a few of the organisms known to live in ponds. Leeuwenhoek died on August 26, 1723, at the age of almost 91.

During the eighteenth and nineteenth centuries, and right up to the present day, the microscope has been improved continuously. Hundreds of different designs have been tried. The single lens microscope began and ended with Leeuwenhoek. All following instruments were based on the principle of the compound microscope. However, mirrors and diaphragms were added, and condensers of different designs were invented to concentrate the light. Light sources were also greatly improved. Microscopical observations began using light from windows, sunlight, or candlelight. Hooke's illustration of his own microscope shows that he used an oil lamp, and that he was so sophisticated as to put a waterfilled vessel in front of the flame in order to concentrate the light (see Figure 1-1).

In an effort to see more and more detail in the specimen, objectives and oculars became more and more complex. An optical science developed gradually, with tremendous achievements along the way. One of these achievements is the phase contrast microscope, invented by Frits Zernicke (1888–1966), a Dutch professor of physics, who received the Nobel Prize for it in 1953. Many candidates for study under the microscope, such as cells or tissues, are transparent and colorless and accordingly have to be killed and stained before they are suitable for observation. Phase contrast permits the microscopist to see some structures in the living, unstained material. Another advance during the last 20 years has been achieved with the development of the so-called interference microscopes. In these modern (and expensive) instruments, the entering light, which consists of bundles of single rays, are deliberately made to interfere with each other in a controlled way. The result is an image that provides additional information. How this is achieved will be explained in Chapter 3 when the various possibilities to illuminate a specimen are discussed. Several systems already exist, the best known being those named after Nomarski and Jamin-Lebedeff.

Yet with all these improvements, the light microscope has its limitations. Light is a form of energy that travels in waves of various lengths according to its color. The size of the wavelengths used to be measured in angstrom units, which are named after the Swedish physicist, Anders J. Ångstrom (1814–1874). One angstrom (symbolized Å) is 1/10,000 of a micron. One micron (symbolized μ, pronounced mu), now called micrometer (symbolized μm), is 1/1000 of a millimeter. The modern international unit for wavelength measurement is the nanometer (nm), also known as millimicron (mμ). It is equal to a thousand microns or micrometers. In other words, one nanometer is one millionth of a millimeter. On a photograph, the magnification is often expressed by a bar, ⊢10 μ⊣, with or without the two small vertical marks. It means that any structure or area in the photograph has the indicated size.

Violet has the smallest wavelength of visible light, between 400 and 455 nm, depending on the shade. The smaller the wavelength, the more detail can be seen through the microscope. The limit of detectable detail depends on the size of the detail itself; if it is smaller than the wavelength with which it is illuminated, it cannot be seen. For example, if a cluster of viruses is viewed through the light microscope, these particles cannot be observed as consisting of separate entities because they are smaller than the light wave to which they are exposed, even to violet light.

There is light below violet with wavelengths of 400 to 200nm. This light is called ultraviolet or "black" because our eyes cannot perceive it. This wavelength can, however, be very useful in research. Some materials, when exposed to ultraviolet (UV) light, fluoresce; they emit light of longer wavelength, thus becoming visible and revealing part of their micromorphology. This important technique is called fluorescence microscopy.

In addition to UV illumination, another kind of radiation can overcome the limitations of the light microscope: the electron beam. This discovery led to the development of the electron microscope, which increased magnifying power a hundredfold and achieved a resolution of about 0.5 nm.

If an electrical discharge takes place between two electrodes inside a tube from which air has been removed, rays are produced at the cathode end of the tube. These are called cathode rays and consist of electrons. They travel toward the anode with great speed. Though invisible, electrons act very much like light rays. They travel in a straight line but can be deflected and focused by magnets, exactly as light rays are refracted when passing through lenses. If they hit certain crystalline substances, they cause them to emit light and to become visible on a fluorescent screen.

If such an electron stream passes through the thin section of a cell, the structural detail of the section can be recorded. The dense parts of the specimen let fewer electrons pass through than do the more transparent parts of the specimen. As a result, an image is formed on the viewing screen or on a photographic plate.

The new knowledge gained by use of the electron microscope is, of course, tremendous. For the naturalist, however, it is of minor interest whether a membrane of a cell consists of one or two layers, or whether minute bodies like mitochondria are shaped like rods or granular. For a micronaut, the range from low power to a magnification of about 400 × is the most exciting and has generally been retained in the illustrations throughout the book.

2
Your Microscope

Since the invention of the microscope, so many kinds of instruments have been developed that it has become a hobby to collect antique varieties. All of these varieties, except the very oldest that originated in the seventeenth century, have one thing in common: a mirror to direct light from a separate source into the optical components of the instrument. During the last two or three decades, however, microscope design has greatly changed. The mirror has been eliminated and the light, instead, is built into the base of the microscope for convenience. The instrument can be plugged into an electrical outlet such as a household appliance without a separate illuminator. The problem of coordinating the lamp with the instrument in order to achieve even illumination is now, at least to a certain extent, prearranged by the manufacturer.

Two or three years ago, the Zeiss company in West Germany even developed an automatic microscope, its KM model. If the observer wishes to change the magnification, for instance from 100× to 400× ("×," in this context, means "times"), he or she would normally have to readjust the optical set-up. In the KM, this readjustment is done automatically. This newest achievement is no doubt a useful improvement for the busy pathologist or technician who may have to scan hundreds of preparations a day, but is not necessary for the naturalist. Besides, "old-fashioned" microscopes are still being made, and they are less expensive. The beginner may also prefer to acquire a used microscope in order to reduce the initial cost of the investment.

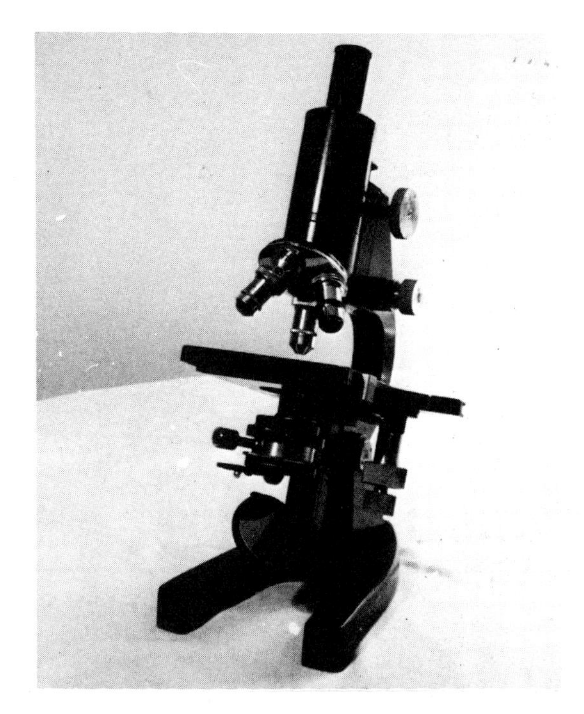

FIGURE 2-1. The author's first microscope.

It is for these reasons that a microscope, vintage 1945, my first, is selected as a teaching model in these pages. A photograph of this instrument is provided in Figure 2-1. It represents the minimum requirements of a microscope necessary to take the traveler into the "invisible."

A microscope should have (see Figure 2-1) a *solid stand*. The upper part consists of the *tube* with the *objectives* on one end, the *eyepiece*, or *ocular*, on the other. Some microscopes are equipped with a revolving *nosepiece* into which objectives of different magnifications can be screwed. Turning the nosepiece makes it possible to change quickly from one magnification to the other. The tube is connected to a *rack-and-pinion mechanism* for focusing. The large knob is for the coarse adjustment of the magnification, the small one for the fine adjustment. When focusing, the whole tube, with the lenses in place, moves up or down. If a large-format camera attachment is put on the microscope, the weight may prove to be too much of a burden; the focus can be affected if the delicate fine adjustment does not hold up. In modern microscopes, the upper part of the stand is rigid, and focusing is done by lowering or raising the stage with the specimen on it. This modification is a definite improvement.

Microscope tubes have a length of either 160 mm or 170 mm. It is important to know the tube size because objectives are computed for specific tube lengths. For instance, objectives made by the American Optical Company, Bausch & Lomb, Nikon, or Zeiss are computed for 160-mm tubes, while those made by Leitz are computed for 170-mm tubes. The tube length, if not known, is measured from the milled shoulder of an objective to the upper end of the tube.

The lower part of the stand houses the stage, the condenser, and the mirror. The *stage* carries the specimen, which is usually mounted on a *microslide,* or glass plate. The microslide's standard sizes are 1×3 or 2×3 inches. The slide is held in place by clamps. When searching for a suitable area of the specimen, the slide is moved back and forth by hand. Moving is often difficult, especially at observations with high magnifications. For this reason, a *mechanical stage* is often built into the stage so that the specimen can be moved short distances in the north-south or east-west directions by the use of two knobs. Separate, attachable mechanical stages are also available.

The *condenser* concentrates the light source. Either for observation or photomicrography, it is of great importance in achieving even illumination of the field of view. The condenser has two or sometimes three lens elements. These elements can be used in combination, or one or two of the upper elements can be removed. The choice of the lens combination to use depends on the intended magnification. For magnifications below 100× (10×objective times 10×eyepiece), for example, the top lens has to be unscrewed or swung out, according to the mode provided by the manufacturer. The condenser is equipped with an *iris diaphragm,* called the *aperture diaphragm.* It is operated with a lever for opening and closing. This diaphragm is of great importance in controlling the quality of the resulting image, as we shall see. Located below the diaphragm, is a swing-out *filter holder.*

The *mirror* usually has a flat surface on one side and a concave one on the other. It is mounted so that it can be turned around and also shifted sideways. If a substage condenser is provided, only the flat surface should be used. The concave side throws a scrambled light bundle from the light source into the condenser and ruins the illumination. The concave mirror is a relic from the times when microscopes did not have substage condensers. However, the concave mirror can come in handy even today for very low power when the field of view cannot be filled with light. In such a case, the condenser has to be removed and the concave side of the mirror turned in to fill the field of view with light.

Eliminating the mirror has reduced the versatility of the modern microscope. A mirror can be used to the advantage of the microscopical image by slightly tilting it so the light rays enter the condenser at an angle, pro-

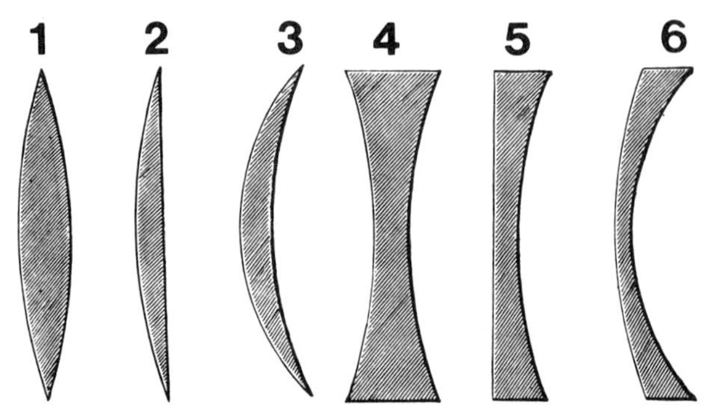

FIGURE 2-2. Types of lenses commonly used on optical instruments such as a microscope; 1 to 3—bi-convex, plano-convex, converging meniscus lens; 4 to 6—bi-concave, plano-concave, diverging meniscus lens.

viding an oblique illumination that gives a pleasing three-dimensional effect.

The heart and soul of the microscope is the *objective*, so called because it is the lens nearest to the object. The ocular, the condenser, and the lamp are its accessories and can improve the image, but only within the limits of the fundamental performance of the objective. To understand its function, a few basic facts need to be explained.

Optical science distinguishes six elementary types of lenses, which are shown in Figure 2-2. Numbers 1 through 3 are called positive lenses; numbers 4 through 6 are called negative lenses. The positive lenses are thicker in the middle than at the periphery, while the negative lenses are thicker at the periphery than at the center. Numbers 1 through 3 are named, in the order of the diagram, bi-convex, plano-convex and concave-convex or meniscus-converging lenses; numbers 4 through 6 are bi-concave, plano-concave, and concave-convex or meniscus-diverging lenses.

When light rays traveling through air strike a surface of transparent matter of greater density—for example, glass, plastic, or crystal—the rays are refracted. The refractive index of a medium is expressed with the letter "n." The refractive index for air is 1.00; for ordinary glass, it is 1.52. Therefore, if light propagating in air in a straight line hits a lens, its path is deflected. The degree of refraction not only depends on the material the lens is made of, but on the shape of the lens and the wavelength of the light. The more rounded a lens, the greater the deflection.

As youngsters, most of us at one time used a magnifying glass to burn holes in a sheet of paper. What happened during that exercise is illustrated in Figure 2-3. In this case, light rays from the sun, which are parallel, go through a bi-convex magnifying glass and converge on P, the primary focal point. The distance from the focal point to the center of the lens is the focal

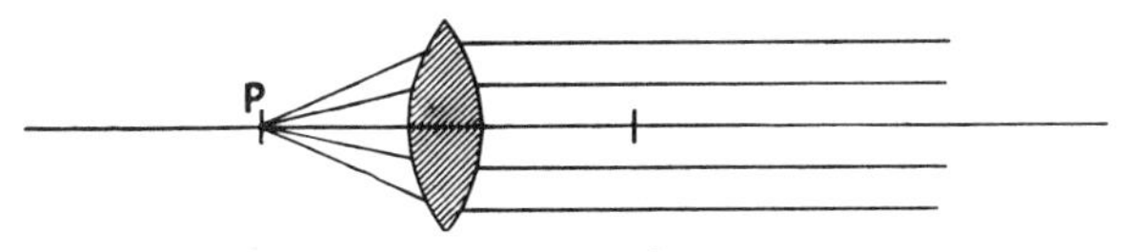

FIGURE 2-3. Light at the principal focus *P* makes the refracted rays parallel or, vice versa, parallel rays, as from the sun, converge at the principal focus of a bi-convex lens.

length of this lens. Vice versa, if we have a point-shaped light source at P and let its rays pass through the lens at the proper distance from P, the emerging rays will be parallel.

If we now move the light source back and forth, closer or farther away from the lens, the focal points change. If the light source is placed beyond the principal focus, the rays converge beyond the principal focus on the other side of the lens (see Figure 2-4). However, if the light is placed between the lens and its principal focus, the rays diverge on the other side of the lens. They do not come to a focus. Yet if they are traced backward, as shown with dotted lines in Figure 2-5, they converge theoretically at a point on the axis called the virtual or conjugate focus. A similar effect is achieved in the case of a bi-concave lens with parallel rays entering (see Figure 2-6).

The first objectives of early micronauts such as Janssen, Leeuwenhoek, and Hooke were bi-convex lenses with two rounded surfaces. These lenses gave a magnified, but poor image. Hooke recognized the inferiority

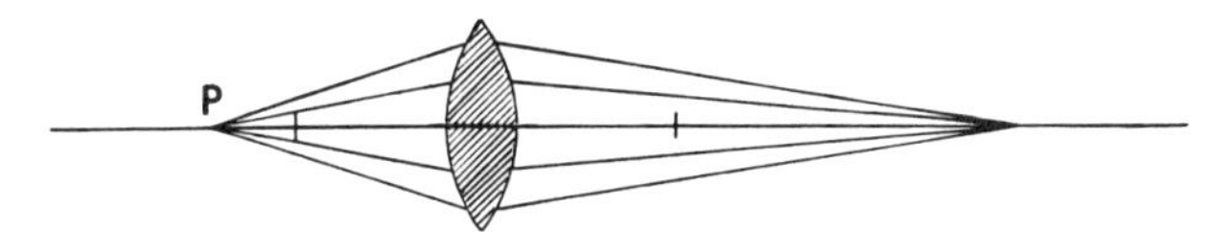

FIGURE 2-4. Light placed beyond the principal focus causes rays to converge beyond the principal focus on the other side of the lens.

FIGURE 2-5. Rays converge if a light is placed between a lens and its principal focus. The focus of divergent rays is called virtual.

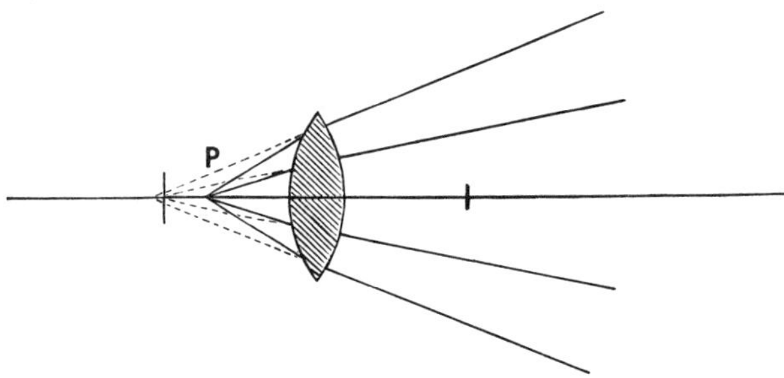

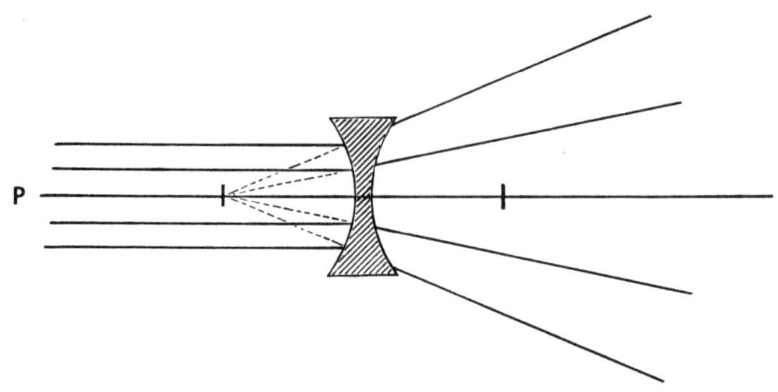

FIGURE 2-6. Parallel rays entering a biconcave lens diverge on the other side of the lens and do not come to a focus.

FIGURE 2-2 to 2-6. *From* William P. Carpenter, *The Microscope and Its Revelations,*
8th edition. (Philadelphia, Pa.: P. Blakiston's Son & Co., 1901), pp. 13–14.

of these objectives and even the reasons for it. In the preface of his "Micrographia," he wrote:

> *. . . for the fewer the Refractions are, the more bright and clear the Object appears. And therefore, 'tis not to be doubted but could we make a Microscope to have only one Refraction it would,* ceteris paribus [*other factors being equal*] *far excel any other that had a great number.*

The shortcomings Hooke complained about are the aberrations his lenses suffered from. The drawings in Figures 2-2 through 2-4, presenting the light path through a bi-convex lens are, in fact, idealized in that they disregard, for the purpose of clarity, the irregularities that have troubled lens makers for centuries. These irregularities are the so-called *spherical and chromatic aberrations.*

The spherical aberration is caused by the fact that the rays passing through the lens actually do not come to a focus at one point. Rays entering the bi-convex lens at its periphery converge at a point nearer the lens, while those entering closer to the center converge farther away from the lens. The effect is a fuzzy image. A ray going through the center is not affected (see Figure 2-7).

The chromatic aberration is brought about by another law of optics: White light consists of different colors. Rays of different colors have different wavelengths. Blue light waves, for instance, are shorter than red light waves. The shorter ones are brought to a focus nearer, the longer ones farther away from the lens, causing an additional decrease in sharpness (see Figure 2-8).

The chromatic aberration can be minimized by using monochromatic light. A green filter, for example, cuts out all colors except green. To put

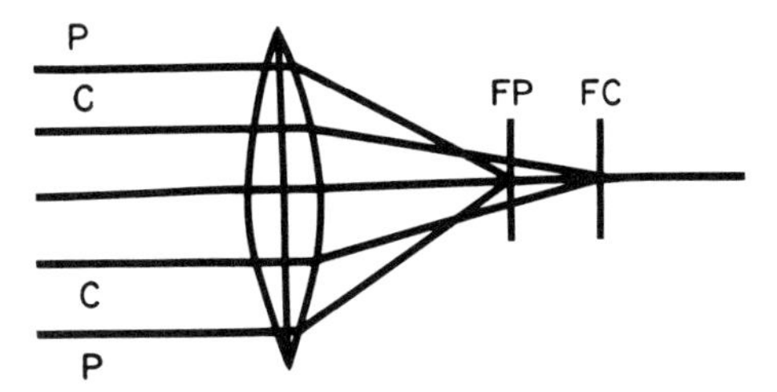

FIGURE 2-7. *Spherical Aberration.* Peripheral rays *P* entering a biconvex lens come to a focus at *FP,* those rays more to the center *C* at *FC.*

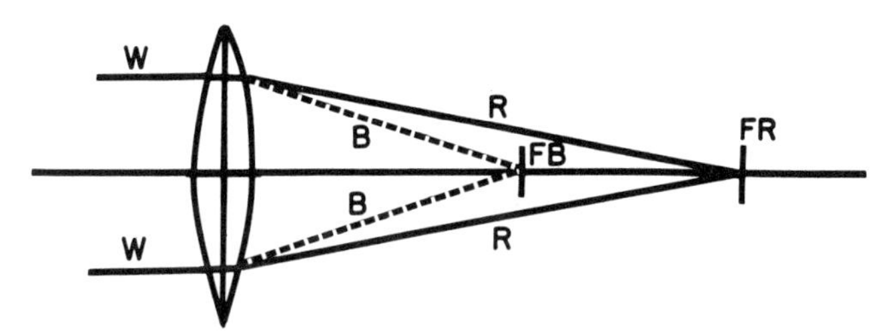

FIGURE 2-8. *Chromatic Aberration.* White light entering a bi-convex lens is deflected at different angles. Blue rays *B* come to a focus nearer the lens at FB; red rays *R* further away from the lens. Intermediate colors have focal points in between.

such a filter into the light path is a common and effective makeshift technique, although it has the disadvantage of diminishing the intensity of the light.

To eliminate these aberrations optically was the unending effort of lens makers for more than 300 years. In the course of these efforts, three types of objectives have been created: the achromats, the fluorites, and the apochromats. The achromats are corrected for spherical aberration, while the fluorites and the apochromats are corrected for spherical and partly for chromatic aberrations. These corrections are possible by making objectives consisting not only of a single lens, but of a combination of several lens elements. A high-powered apochromat may have up to 11 components combining many different lens shapes, as shown in Figure 2-2. Apochromat objectives also require specially designed eyepieces called compensating eyepieces in order to improve the image. Some objectives are made of different qualities of glass, each one having slightly different refractive indices with respect to a specific wavelength of light. The same principle is incorporated in the fluorites, in which one or more of the lens elements consist of the mineral of this name. Some apochromats are even corrected for three different colors.

There is another distinction among objectives: those that can be used dry and those that must be immersed in transparent liquids, such as oil, glycerine, water, or another clear, colorless liquid of specific refractive index. All low-power lenses and most medium-power lenses are dry objectives. Some medium-power lenses and all high-power objectives are immersion objectives.[1]

The reason why some lenses must be immersed is again prompted by the need to correct the adverse refractive phenomena that nature built into the basic optical laws. As we have seen, light passing from air into glass is deflected. The same rule exists in the reverse situation. For example, rays traveling through the cover glass (the thin platelet usually put over the embedded specimen as a protective cover) enter the air before entering the objective and are also deflected. This refraction is especially disturbing at high magnifications.

Figure 2-9 graphically shows the effect on light rays passing from glass into air. From the bundle of rays, some enter the objective, others are excluded from it. Not only is light lost, but the quality of the image suffers too. If it were possible to fill the air space between the cover glass and the objective with a matter that had the same refractive index as glass, then the adverse effect of the air/glass passage could be eliminated. This was the problem that physicists in the 1870s solved by developing the oil immersion objective. They found that cedarwood oil has very nearly the same refractive index as the cover glass. Physicists designed lenses that could be immersed in a drop of oil between a cover glass and the objective. For uncovered aqueous specimens, water or glycerine is the immersion medium. The effect of such lenses on the light path is shown in Figure 2-10. The air space is filled with oil, the immersion liquid, and hence, the rays are not dispersed. The light-gathering power of the lens is increased and with it the resolving power.

FIGURE 2-9. From a bundle of light rays passing through air, only rays *1* and *2* enter the objective.

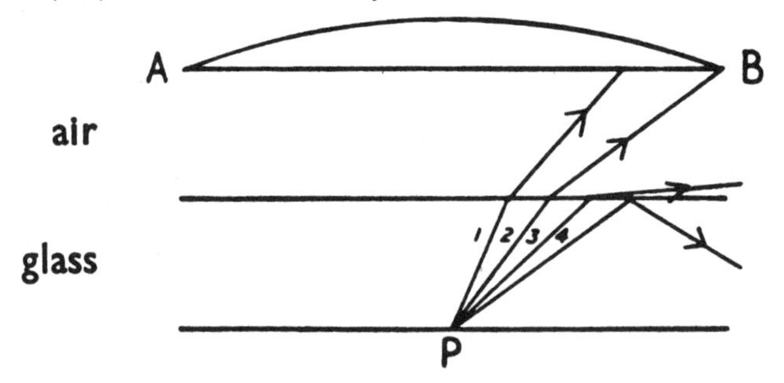

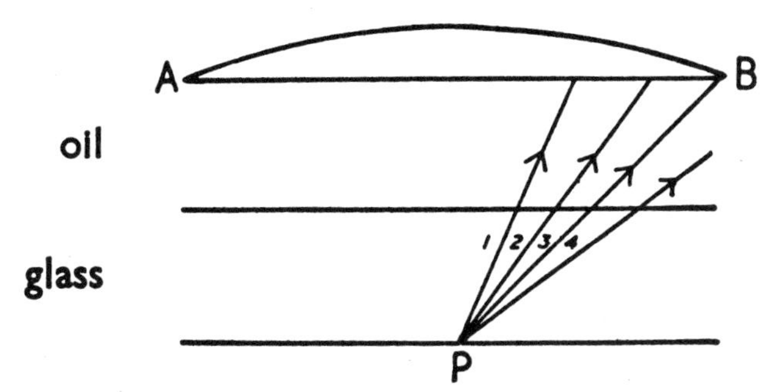

FIGURE 2-10. From a bundle of light rays passing through oil, rays *1* and *3* enter the objective.

The light on its way from the lamp to the ocular encounters many obstacles, many refractions: the condenser, the slide, the mounting medium in which the specimen may be embedded (or mounted in air), the cover glass, and finally, the air space below the objective. For this reason also, the condenser should be oiled to eliminate the air gap between the top lens of the condenser and the slide on which the specimen rests. The specimen itself should be mounted in a medium with the same or nearly the same refractive index as glass, such as canada balsam or one of the more modern, fast-drying resins. A great number of these are available.

There are still more problems for the objective manufacturer to solve. One such problem is the *flatness of the field*. Older objectives provide an image in which only the center of the field of view is sharp, while the sharpness gradually deteriorates toward the periphery. This phenomenon is not to be confused with spherical aberration, which affects the sharpness of the whole image, not only the periphery. Flatness is a problem that is more pronounced the higher the magnification. This problem is especially annoying in photomicrography because it severely restricts the useful extent of the image. With a 10× achromatic objective, about 80 percent of the image is sharp; with a 100× objective, only about 50 percent is sharp. In modern objectives, this problem has been solved. Such lenses are labeled as plan-achromats or plan-apochromats.

Another refinement of some objectives is the *cover glass correction collar*. Most permanently mounted specimens, such as histological sections, are protected by cover glasses. The thickness of these cover glasses varies. The standard thickness of the recommended grade number 1.5 cover glass is 0.18 mm. Often, the thickness of the cover glass on commercially available preparations is not known. Yet it makes a difference for critical work whether this thickness is 0.16, 0.17, 0.18 mm or thicker. The correction is done by turning the collar while, at the same time, visually determining at which position the image is sharpest. The correction is achieved by varying the distance between the front lens and the back lens of such an objective.

All objectives have more or less elaborate engravings. What is their meaning? One of the finest objectives, made by Carl Zeiss, West Germany, bears the following "hieroglyphs": "Ph3 Planapo 100/1.3 oel 160/-." "Ph" stands for phase contrast, to be used with the #3 condenser diaphragm. (Phase contrast is a special kind of illumination that will be discussed later.) "Planapo" stands for an apochromat corrected for flat field. The number "100" means that the lens magnifies one hundred times. The number "1.3" indicates the numerical aperture. The "oel" we can guess. The number "160" specifies the tube length for which the lens was computed.

The most important information among these engravings is the numerical aperture, usually abbreviated "N.A." It is based on a formula by the very famous German physicist, Ernst Abbe, who developed it mathematically at the end of the last century. The N.A. is determined by the manufacturer of the instrument. Workers with the microscope rarely, if ever, have a need to compute it themselves. If such a need arises, such as for an old lens not being marked with this information, special instruments, called apertometers, can provide assistance. It is important, however, to know what the N.A. means: It indicates the light-gathering and the resolving power of the lens. The N.A. of a microscope objective is roughly related to the f-number of a camera lens by the formula:

$$f = \frac{1}{2\ \text{N.A.}}$$

The formula for the numerical aperture is: N.A. $= n$ sine u, in which "n" is the refractive index the light passes on its way from the cover glass to the front lens of the objective. For a dry lens, the medium is air, and n is 1; for an oil immersion lens, n is 1.52; and for a water immersion lens, n is 1.33. The symbol "u" is half of the cone of light the lens can admit. The full angle is called the angular aperture.

Figure 2-11 shows graphically the factors that determine the angular aperture and through it, the N.A. Two dimensions must be known:

1. the diameter of the front lens of the objective, line A to B,
2. and the focal length of the objective, line P to C.

From these measurements, the angular aperture can be derived. As to the sine, "u", this value is listed in natural sine tables contained in physics textbooks and handbooks. The one shown here is taken from Carpenter's *The Microscope and Its Revelations* (Philadelphia, PA: P. Blakiston's Son & Co), 8th edition, 1901, Appendix A.

The exact dimensions of items 1 and 2 are known only to the manufacturer of the microscope. In the drawing (see Figure 2-11), the angular

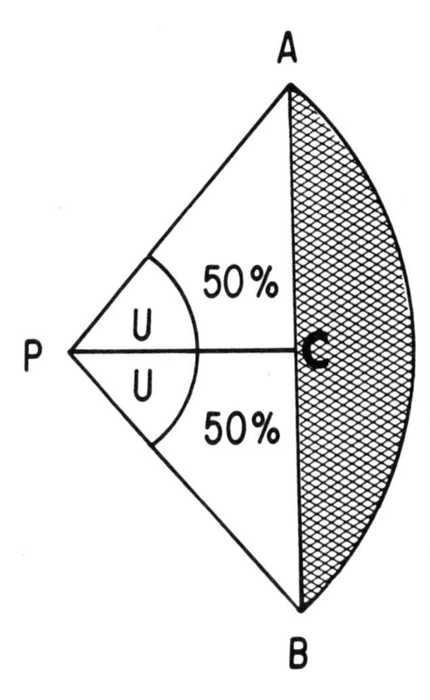

FIGURE 2-11. The numerical aperture is derived from the angular aperture, outlined by the cone of light *PA–PB*. The sinus of *u*, half of the angular aperture, multiplied by the refractive index *n* of the medium between the cover glass and the objective determines the N.A.

FIGURES 2-9 to 2-11. *Courtesy* R. Barer, "Lecture notes on the use of the Microscope," Oxford, England: Blackwell Scientific Publications Ltd., 1956.

aperture is assumed to be 117.6°. Half of it *(u)* would be 58.8°, the sine of which is 0.8549. In figures, the formula "*n* sine *u*" would read:

for a dry lens:	N.A. = 1 × 0.8549 = 0.855 (0.85);
for an oil lens:	N.A. = 1.52 × 0.8549 = 1.299 (1.3);
for a water lens:	N.A. = 1.33 × 0.8549 = 1.137 (1.1).

It follows that the N.A. increases with the refractive index of the medium and the angular aperture. You may ask why this angle is not further increased in order to produce lenses with still higher N.A.s. This has been tried in the past, but such lenses are not feasible. The higher the angular aperture, the closer the front lens has to be to the specimen under observation. The increase of the angle necessarily shortens the working distance between the objective and the cover glass until, at the theoretical extreme of 180°, the working distance becomes zero. In the 100×/1.3 apochromat used here as an example, the working distance is already as small as 0.15 mm.

At this point, an explanation of the expression *working distance* is in order. A microscope has often been compared to an extended, and by far more complicated reading glass. This comparison is correct to the extent that both work according to the same optical laws. A reading glass is only slightly curved, though enough for its purpose of enlarging print to a small extent. It must, however, be held at a definite position in order to produce a sharp image. The distance of the glass from the object is called the working distance. The higher the magnification, the shorter the working distance. It finally gets down to fractions of a millimeter for oil immersion lenses. Indeed, the working distance can be less than the thickness of a cover glass. This is one of the reasons why an immersion objective may be used without a cover glass.

The N.A. is the most important indicator for the quality of an objective. It not only determines the "speed" of a lens (to borrow an equivalent from the camera lens), but also relates to its resolving power. Every microscopist will be curious to know in micrometers (1/1000 mm) the utmost performance of his or her oil immersion lens or, for that matter, any other lens. The resolution—that is, the ability of an objective to form distinguishable images of two objects separated by a small distance—can be figured out by a simple formula that takes two important aspects into account: the wavelength of the light employed and twice the N.A., assuming the most favorable adjustment of the condenser aperture. White light consists of all wavelengths of the visible spectrum, from violet to red. A particular wavelength is expressed by the symbol λ (lambda), the Greek letter for "l," the size of which is defined in nanometers (nm) (see page 9). The formula for resolution (d) is:

$$d = \frac{\lambda}{2 \times \text{N.A.}}$$

Accordingly, a 100/1.3 apochromat should be able to resolve, with violet light of 425 nm, two cocci only 163 nm apart:

$$d = \frac{425 \text{ nm}}{2 \times 1.3} = 163 \text{ nm or } 0.163\ \mu$$

A 40×/0.65 apochromat, illuminated with green light of 513 nm wavelength should show two particles 395 nm as separate units:

$$d = \frac{513 \text{ nm}}{2 \times 0.65} = 395 \text{ nm or } 0.395\ \mu$$

The N.A. is not only of theoretical interest, but also has a practical application: It makes it possible to determine the upper limit of feasible magnification through the microscope. The magnification under a given optical set-up is the power of the objective multiplied by the power of the eyepiece.

Yet there is a limit which, if surpassed, is "empty magnification." This would come about, for instance, if a 100×objective were combined with a 20× eyepiece, giving a total of 2000×. The general rule is not to surpass the magnification by 1000 times the N.A. of the objective. A 100/1.3 objective would reach this limit at 1.300×, when combined with a 12.5× ocular. A 15 or 16×ocular might still be acceptable, depending on the quality of the preparation, such as a well-stained bacterial smear. A 20×eyepiece would definitely result in a fuzzy photomicrograph. However, there may be circumstances in which some empty magnification is justified.

The range of magnifications offered by the industry extends from 1× to 100×, with about 17 intermediate ×-values in between. For the beginning, only four objectives are recommended: the 3.5 or 4×; the 10×, 20×, or 25×; and the 40× or 45× objectives. Immersion lenses could be considered much later. Such objectives can be combined with 6×, 10×, and 15 or 16× oculars.

The second component of a compound microscope is the *eyepiece*, or *ocular*. There are two basic types, the Huygenian and the Ramsden oculars. For the Huygenians, optically advanced compensating eyepieces are available. These eyepieces are designed to improve the flatness of the field and are therefore important for photomicrographical purposes.

An eyepiece consists of three elements: a field lens, the one nearest to the objective; the eye lens, the one nearest to the observer's eye; and a fixed diaphragm. In the Huygenian eyepiece, the diaphragm is located between the two lenses. In the Ramsden eyepiece, the diaphragm is below the field lens; it is better corrected for aberrations. Therefore, the Ramsden type is preferred for higher power and for special situations.

The magnifying power of an eyepiece is marked on the eye lens. If the engraving reads "Periplan" or "KPL," the information indicates a compensating ocular. Some old eyepieces sleeping in pawn shops carry the marking "4" or "5," without an ×. This is an antiquated way of identifying the magnifying power of the lens. The unsuspecting buyer may assume that these are low power glasses. Actually, the number 4 indicates a 10×, number 5 a 12×, and number 6 a 15× ocular.

When deciding on the best objective-eyepiece combination to achieve a desired magnification, the microscopist should keep in mind that the primary enlargement of the objective is always the more useful. If, for instance, a magnification of 250× is needed, it can be achieved by using either a 10× objective with a 25× ocular, or by using a 25× objective with a 10× ocular. The latter combination will give the better image.

Another essential part of the microscope is the *light source*. There are several different sources of light. We can put the microscope in front of a window; daylight was used for illumination until the end of the ninteenth century and the advent of electricity. A gooseneck desk lamp with a soft-white bulb aimed at the microscope mirror is only a makeshift light source and is useful only for casual observations. For serious observation and for photomicrography, such a light source is insufficient.

A proper microscope illuminator must have:

1. a small, intense light source, such as a narrow tungsten filament or tungsten ribbon;
2. a collecting lens;
3. a diaphragm;
4. a means of centering the bulb in its housing with reference to the collecting lens, and to focus the filament to various parts of the microscope;
5. a filter holder in front of the diaphragm.

These conditions are fulfilled by the Spencer lamp, which is made by the American Optical Company (see Figure 2-12). This lamp has the advantage of working on household outlets without needing a special transformer. It uses a GE 100-watt, 120-volt bulb with a filament, as shown in Figures 2-13 through 2-15. Its catalogue number is FG-1635-X. There are various other types of microscope lamps available as well. Many of them, however, are low-voltage lamps that not only need transformers, but also special bulbs.

Setting Up the Microscope

We now turn to the question of how to set up a microscope by coordinating all of its components and the illuminator. The first element to deal with will be the lamp, for which the Spencer is selected (as an example to be used in our demonstration).

When using the lamp for the first time, or replacing the bulb, first center the bulb. Centering is done by projecting the filament image, with the diaphragm closed, on a screen or a uniform area of the wall at a distance of about three feet. Most likely, the filament will be out of focus (see Figure 2-13). By loosening the bulb socket, then turning and moving it up and down, it must be placed so that all the four sections of the filament show in sharp focus (see Figure 2-14).

After this, the lamp is put in front of the microscope at a distance of about eight to ten inches. With the lamp diaphragm still closed, aim the light at the flat surface of the mirror so the filament is refocused to throw a sharp image on the mirror's center (see Figure 2-15). To facilitate this, a piece of white paper or cardboard can be loosely taped over the mirror. It is now important to center the incoming light in such a way that its path goes vertically through the microscope tube. For this purpose, the condenser, the objective, and the eyepiece are removed. In place of the ocular, a pinhole eyepiece is inserted into the tube. This gadget looks like an ocu-

FIGURE 2-12. The Spencer microscope lamp.

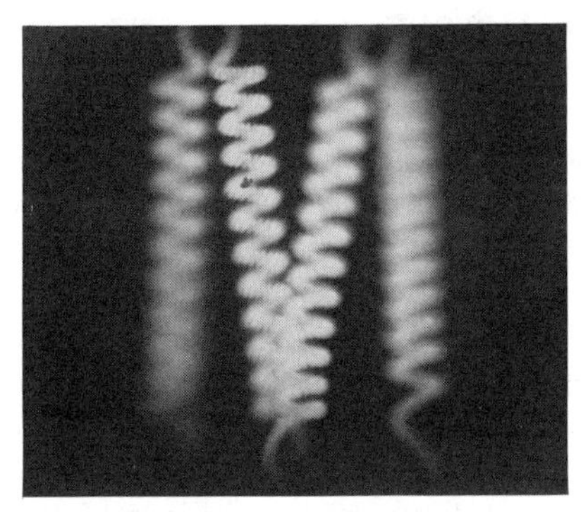

FIGURE 2-13. By centering the lamp filament against the wall, the image is likely to be out of focus.

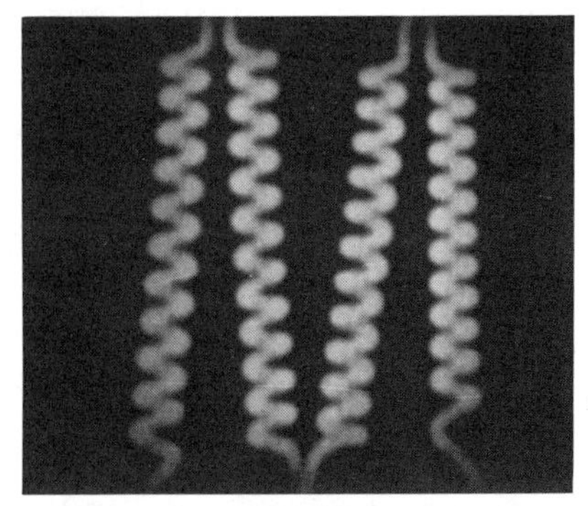

FIGURE 2-14. Image of the filament as it should look after sharpening.

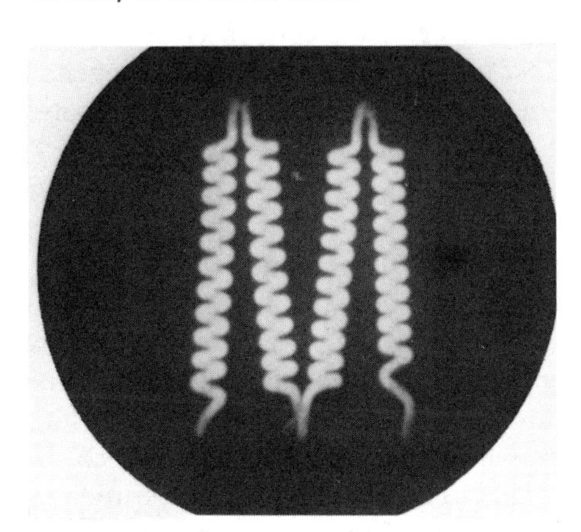

FIGURE 2-15. The lamp filament is focused on the mirror.

FIGURE 2-16. With the optical components removed, the light from the illuminator must now be aligned axially by manipulating the mirror. Looking into the tube with a pinhole eyepiece, the image of the field diaphragm should be in the center of the field of view.

lar, but instead of the lenses it has a tiny hole in the center. The pinhole eyepiece eliminates the intense glare the eye would meet if one were to look directly into the tube.

What is seen through the pinhole is a small, bright spot of light on a dark disc. The spot is the image of the closed field diaphragm on the lamp. The brilliant spot has to be brought into the center by manipulating the mirror. The light beam entering the microscope is now axially aligned with the tube (see Figure 2-16). The objective, eyepiece, and condenser are replaced, the field and aperture diaphragms are temporarily opened, and the specimen is put on the stage and focused.

What follows is an adjustment for what is called Köhler illumination. This method consists of a few important steps to ensure even, well-centered illumination and the elimination of glare. For this, one must close both diaphragms again and focus the lamp filament, this time not on the mirror, but on the closed aperture diaphragm below the condenser. Naturally, it is a little difficult to look at this diaphragm without breaking one's neck. A pocket mirror placed under the microscope mirror (without moving it out of alignment) can make the filament clearly and easily visible (see Figure 2-17).

After this, one should have another look at the specimen without changing the two diaphragms. Again, there will be a bright but fuzzy spot in the field of view, but more to the periphery. This, once more, is the image of the field diaphragm on the lamp. It means that the condenser is not in focus (see Figure 2-18). The focus can be corrected by moving the condenser up or down until the diaphragm is well outlined (see Figure 2-19). Even then, the bright spot will still be at the periphery of the field of view and must be brought to the center (See Figure 2-20). This can be done by making a slight adjustment with the mirror or, if the condenser has centering screws, by turning the screws. Now the field diaphragm can be opened as far as necessary in order to fill the field of view with light—but not more! (see Figure 2-21).

There remains one more very important step to be taken, the proper adjustment of the aperture diaphragm. At this point in the procedure, the aperture diaphragm, also called "iris" or "iris diaphragm," is still closed, actually too much so, and the image does not look its best. There is too much diffraction, an effect light rays undergo if they pass through a narrow slit or a closed diaphragm. Touching the edge of the aperture causes some rays to be deflected, interfering with the otherwise carefully adjusted illumination. The iris, on the other hand, should not be opened too much, or the specimen will be flooded with light. The most favorable position is somewhere in the middle, at about half or three-quarters of the iris diameter. The position can be verified by use of the pinhole cap, with which the back lens of the objective can be inspected. However, even without the pin-

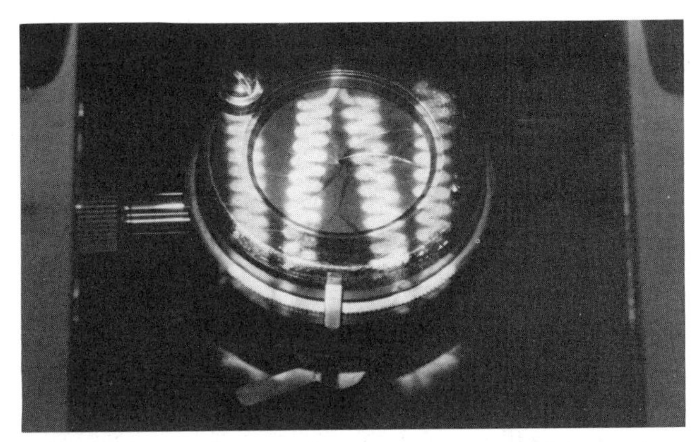

FIGURE 2-17. The filament is focused on the iris of the condenser diaphragm with the help of a pocket mirror placed underneath the microscope mirror.

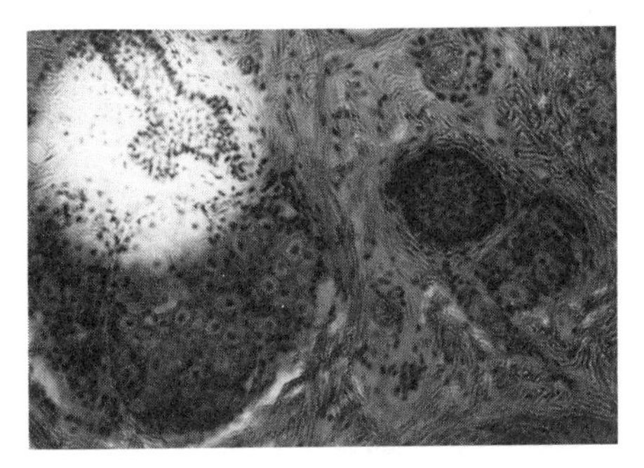

FIGURE 2-18. The field diaphragm is out of focus and also at the periphery, not properly centered.

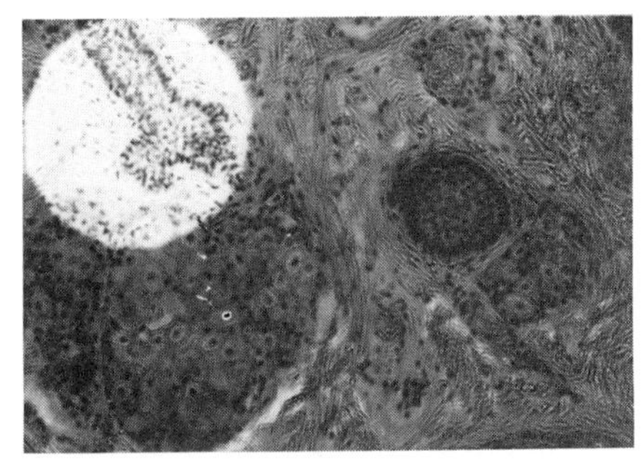

FIGURE 2-19. By moving the condenser up or down, the diaphragm is sharpened.

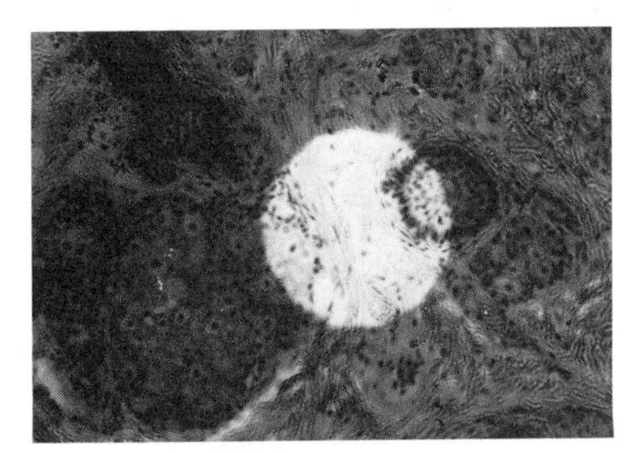

FIGURE 2-20. By slightly adjusting the mirror or centering screw of the condenser, the image of the field diaphragm is brought into the center.

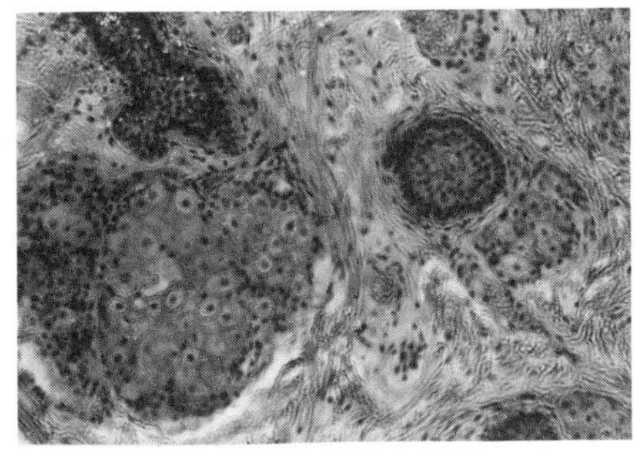

FIGURE 2-21. The field diaphragm can now be opened but only to the extent necessary to illuminate the field of view.

hole cap it is possible to judge the size of the aperture by closing and opening it while at the same time viewing the specimen. One rule must be observed, though, at all times: Do not close the iris because there is too much light. The light intensity must be controlled only by the use of color or neutral density filters.

The Köhler illumination, shown graphically in Figure 2-22, is the established procedure for setting up a microscope for bright field illumination with ordinary transmitted light. The reader may wonder how the Köhler method can be applied to microscopes with a built-in light source, in which a mirror positioned under the condenser for viewing the aperture diaphragm would cut off the light. Actually "Köhlering" is not necessary. The illumination in such microscopes is prearranged by the manufacturers to give even lighting. All of the final adjustments can be done with the condenser: At each change of the objective, the field diaphragm (at the base) has to be refocused in the plane of the specimen and centered; then, as the last step, the aperture iris has to be properly set.

Another method for adjusting the light may be of interest to those who, at least in the beginning, would be satisfied with a less elaborate lamp than the one described for Köhler illumination. This method is called "critical," or "Nelsonian" illumination," after Edward Nelson, the Englishman who first suggested it. Nelson's method originated at a time when microscopists still depended on the oil lamp for illumination. Those old-fashioned kerosene lamps had one advantage as compared with other choices such as sunlight or light from a window or candle: Its flame gave homogeneous light. Nelson's basic idea was to superimpose two images: the image of the specimen and, on the same level, the image of the light source. This idea worked fairly well up to the time when Thomas Edison invented the electric bulb. The bulb's tungsten filament introduced a second image; the image of the object and the image of the bulb appeared in the same focal plane. Although the latter could be eliminated with a diffusion plate, much light was scattered in the form of disturbing glare.

FIGURE 2-22. The principle of Köhler illumination. (*Courtesy* D. Szabo, Medical Colour Photomicrography, Akadémiai Kiado, Budapest, 1967.)

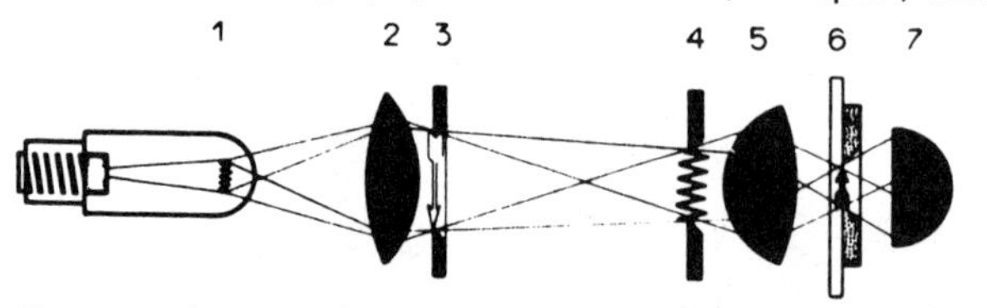

Beam path according to the Köhler principle of illumination.

1. Light source
2. Lamphouse condenser
3. Field diaphragm
4. Aperture diaphragm and image of filament
5. Substage condenser
6. Object plane with image of field diaphragm
7. Objective.

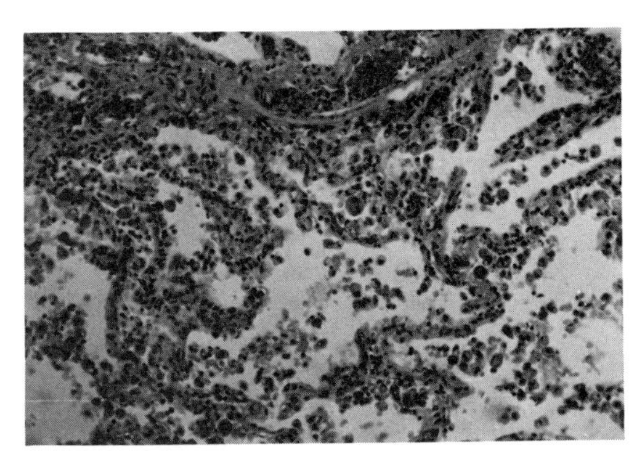

FIGURE 2-23. Photomicrograph of a histological section for the purpose of establishing the focal plane for the intended magnification with a 40 × achromat and an 8× ocular.

Critical illumination is mentioned here because it can be a makeshift for microscopical *observation* with a 100-watt, 120-volt bulb in a gooseneck lamp. Photomicrography is not feasible with this method because the light provides illumination consisting of uncontrolled, scrambled rays going in all directions. When using such a gooseneck lamp, the following steps must be taken.

Put the lamp ten inches in front of the microscope at an angle. Aim the light at the mirror. Remove all the optical elements, objective, eyepiece, and condenser, then insert a pinhole eyepiece into the tube. The mirror should be turned back and forth and sideways until the light from the bulb appears brightest. Then replace the optical components, put a slide on the stage, and focus in order to establish the focal plane for the selected objective (see Figure 2-23). How does one now focus the light in the plane? Moving the condenser up and down, the light source will appear fuzzy in all positions. No filament can be seen, of course, because the milky glass of the bulb obscures it. The solution is to put a black stick-on letter or a small piece of black tape (for instance, a disc cut out with a punch) on the bulb (see Figure 2-24a) where its image can be reflected by the mirror to be seen through the ocular. The letter or tape can be handled with a tweezer. This operation requires some experimentation, but finally a spot will be found on the bulb from which the letter can be aimed correctly and sharpened in the field of view (see Figure 2-24b). With the slide replaced on the stage, one sees a double image (see Figure 2-25) of the specimen with the letter or tape right in the middle. To eliminate the image of the letter, get the condenser slightly out of focus. The careful adjustment of the aperture iris retains its importance in the Nelson procedure.

Figure 2-26 shows the final photomicrograph of a preparation (a section of human lung tissue) taken with critical illumination, but with an

FIGURE 2-24a. An ordinary electric bulb with the stick-on letter "X."

FIGURE 2-24b. Photomicrograph taken for the purpose of focusing the letter "X" that represents the light source in the same plane as the specimen.

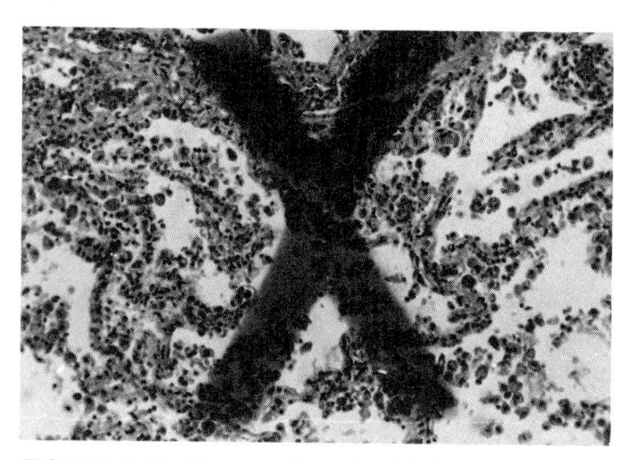

FIGURE 2-25. The resulting double image.

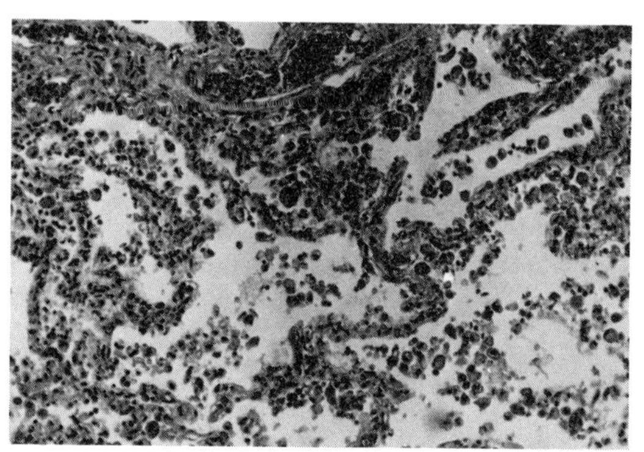

FIGURE 2-26. Final photomicrograph.

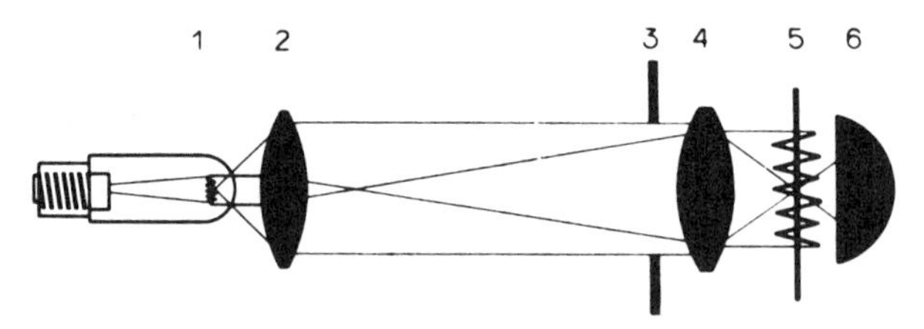

Beam path in critical illumination.

1. Light source
2. Lamphouse condenser
3. Aperture diaphragm
4. Substage condenser
5. Object plane with image of filament
6. Objective.

FIGURE 2-27. The principle of critical illumination. (*Courtesy* D. Szabo, Medical Colour Photomicrography, Akadémiai Kiado, Budapest, 1967.)

unorthodox light source. Using a microscope lamp, this method is a valid alternative to the Köhler method. It is by no means outdated and is still used with a suitable illuminator, such as the ribbon filament lamp made by Bausch and Lomb. In place of a coiled filament, the ribbon filament lamp has a tungsten ribbon, 1×6 mm in size, which gives an even and intense light. Because of the homogeneity of the ribbon, this excellent illuminator can be used without a diffusion glass (ground glass). The principle of critical, or Nelsonian illumination is shown graphically in Figure 2-27.

Minimum Equipment for a Beginner

Before bringing this chapter to a close, the question remains of the minimum equipment the beginner should acquire. The microscope of a beginner should not be too simple and should be one that leaves open the possibility of adding to it later on. The very cheap microscopes, those sold in drug stores or gift shops, are out of the question. They are a waste of money and would only disappoint and discourage the fledgling micronaut. Briefly, your microscope should have the following features:

1. It must have a solid stand, with coarse and fine adjustments for focusing, and a monocular tube. A binocular is a convenience, but is more expensive and would need two matching eyepieces.

2. The microscope should have a revolving nosepiece to hold three or four objectives.

3. The following objectives are recommended: (a) a 3.5 or 4×; (b) a 10×; (c) a 20 or 25×; (d) a 40 or 45×.
   A high-power oil immersion lens (90 or 100×) is less frequently needed and can be acquired much later. The recommended objectives can be achromats.

4. The microscope should be equipped with the following eyepieces: (a) a 6×; (b) a 10×; (c) a 15 or 16×; (d) a pinhole eyepiece for use in alignment.
   If photomicrography is planned, flat-field compensating oculars are recommended, provided the objectives are also plan-achromats to give a flat primary image. Uncorrected oculars may throw a good image from the objective out of focus at the periphery of the field of view.

5. A condenser that can be focused up and down is also a necessity. The condenser must have an aperture diaphragm and a filter holder underneath. For microscopes with built-in illumination, no filter holder is needed. The filters can simply be placed over the light in the base of the microscope. There are many different types of condensers. For the beginner, an Abbe two-lens type condenser is satisfactory. The top lens should be removable.

6. A mechanical stage is optional.
7. A microscope lamp with a collecting lens, a diaphragm, and a means of moving the bulb for focusing is a necessity unless the microscope has a built-in light source.
8. For the photomicrography of living organisms, an electronic flash is indispensible. Along with the electronic flash unit, a light source for observation also will be necessary, because the duration of an electronic flash (1/1000 sec. or less) is too brief to allow viewing the specimen or adjusting the microscope. We shall come back to this problem in the chapter on photomicrography.

3
Special Methods of Illumination

Brightfield Illumination

In Chapter 2, the ordinary method of illumination for ordinary transmitted light, known as "brightfield," was emphasized. Its use results in an image that shows the object in black or grayish tones on a bright background, as shown in Figure 3-1a, a micrograph of *Paramecium caudatum* photographed at 320×. The experienced micronaut can recognize two organelles within this protozoan, the contractile vacuoles, which control the water content of the cell. In the diastole phase shown, the rosette-shaped canals are in the process of being filled with fluid. The contractile phase, systole, eliminates the excess. Some food vacuoles of the animal are also outlined, though weakly.

Brightfield is only one of several ways to illuminate a specimen. Over the years, I have photographed the same species *(Paramecium caudatum)* nine different ways, which will be shown as we go along. These nine results demonstrate the fact that with each kind of illumination, the observer is provided with different information.

Darkfield Illumination

Instead of making the specimen appear on a bright background, this method shows the specimen against a dark background. In this method, an opaque disc of the appropriate size is placed in the filter carrier underneath

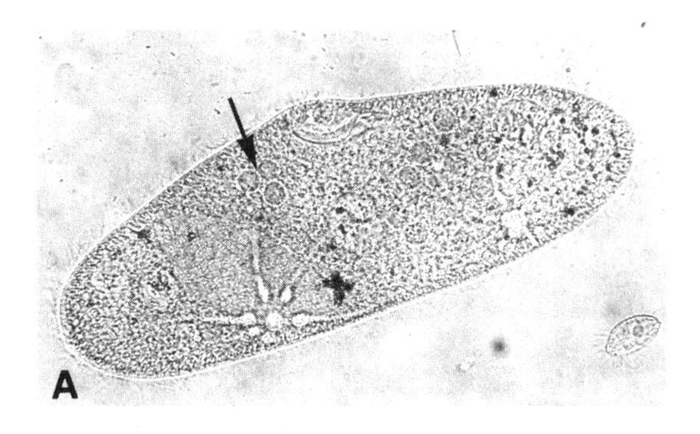

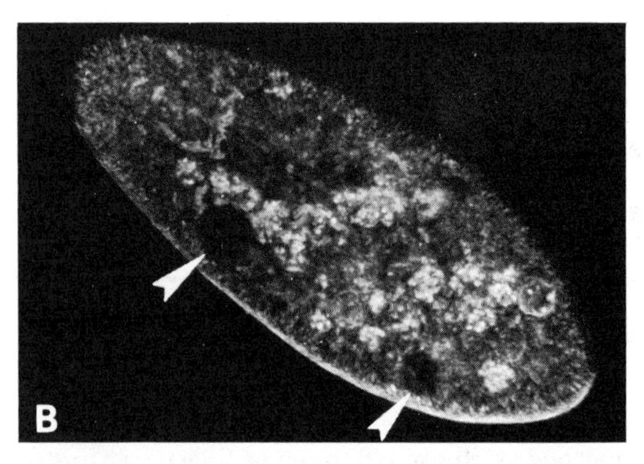

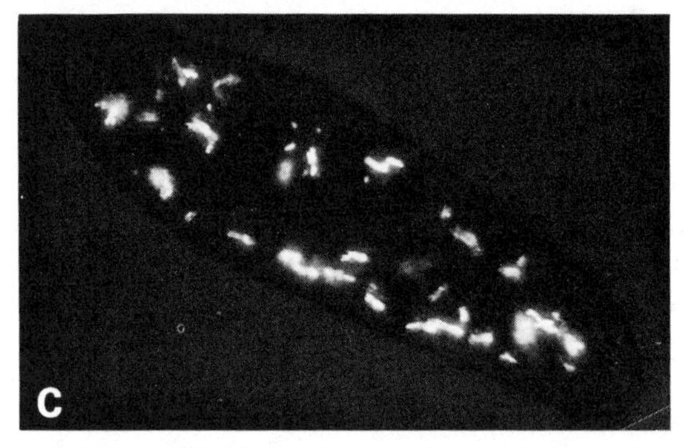

FIGURE 3-1. (a) *Paramecium caudatum* photographed with brightfield illumination. Magnification is 200 X. (b) *Paramecium caudatum* photographed at the same magification with a darkfield condenser, 200 X. (c) *Paramecium caudatum* photographed with polarized light between crossed polars. Only the birefringent crystals of calcium oxalate can be seen. (d) *Paramecium caudatum* photographed with modulation contrast, 200 X. (e) *Paramecium caudatum* photographed with phase contrast, 200 X. (f) *Paramecium caudatum* photographed with differential interference contrast after Nomarski, 200 X.

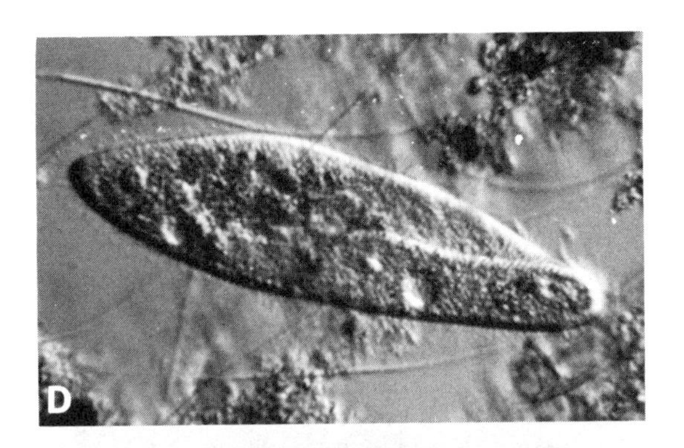
D

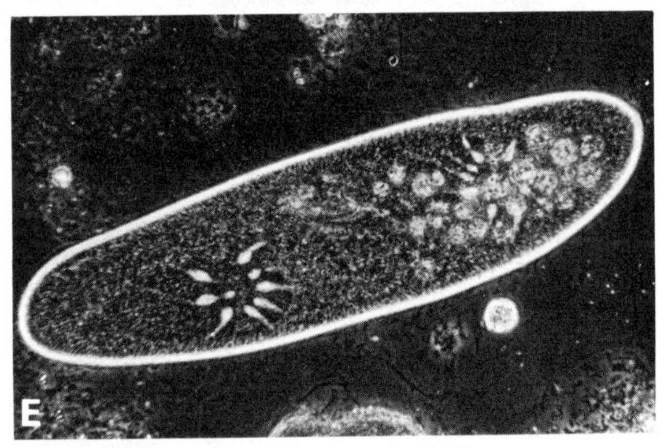
E

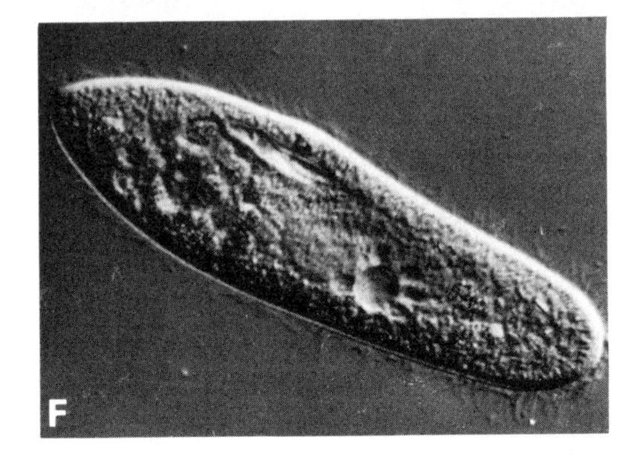
F

FIGURE 3-2. Disc placed below the condenser for darkfield illumination; *S*=stop, *P*=peripheral area.

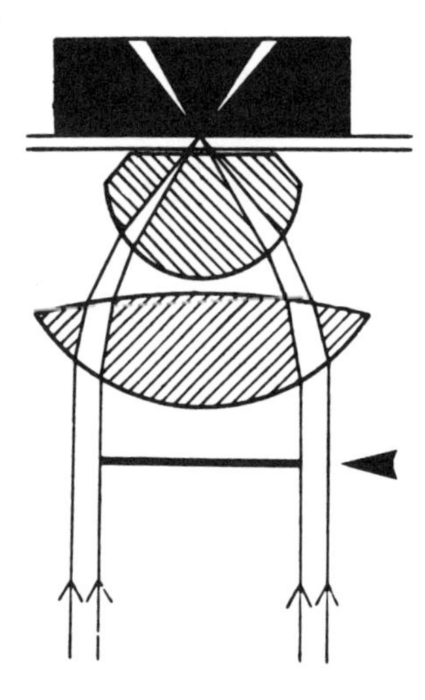

FIGURE 3-3. Path of rays through a 2-lens Abbe condenser (arrow). *From* George Herbert Needham, *The Practical Use of the Microscope,* 1958. (*Courtesy* of Charles C Thomas, Publisher, Springfield, Illinois.)

the condenser. This "stop," or "S" as it is called, has the function of preventing direct light from entering the objective. The specimen is illuminated only by scattered light passing through the periphery ("P") of the condenser (see Figure 3-2).

A great many darkfield condensers have been developed since the Reverend J. G. Reade, an Englishman, devised such an accessory in 1837. These condensers differ according to the number of lens elements they contain and the place where the stop is located. The stop can be underneath, inside, or on the top lens of the condenser. Darkfield condensers are all expensive, however. Fortunately, it is not too difficult to convert a brightfield condenser to a darkfield one. The most practical condenser for this task is the Abbe two-lens type that has a removable top lens. The principle of darkfield illumination is shown in the diagram in Figure 3-3. The stop is indicated with an arrowhead.

To arrange for darkfield lighting, a transparent support for the stop is needed and must be the proper size to fit the microscope's filter carrier. For

FIGURE 3-4. Photomicrographs of the backlens of a 6 X objective. It indicates that the proper diameter of a darkfield stop should be 4.5 mm. See text.

this purpose, a disc of clear plastic such as Plexiglass may be made by a mechanic who is equipped to cut it to size. (Several such discs should be cut, not only to accommodate stops of different sizes, but also because they may come in handy for other methods to be discussed later.) An alternative would be to have glass supports made by an optician, but this would be more costly. Whether made of organic or inorganic glass, the thickness of the discs should not exceed 1 mm.

The filter carriers on microscopes come in different sizes. The Leitz microscope shown in Figure 2-1 takes filters that are 29 mm in diameter. Other microscopes take larger filters: Zeiss, 33 mm; American Optical, 32 mm; Bausch & Lomb, 33.5 mm. The stop supports must be cut accordingly.

With regard to the size of the stop itself, textbooks and papers on the subject indicate that the stop has to be adapted to each objective with which it is to be used. The size of the stop diameter can be easily determined. Begin by focusing a 6× objective and the condenser on the specimen. Then insert a transparent millimeter ruler or a piece of graph paper with millimeter gratings between the filter carrier and the condenser. Remove the specimen and the eyepiece and inspect the backlens of the objective with a pinhole eyepiece. In the backlens, one can count 4½ millimeter lines, the minimum size of a darkfield stop for this objective. (See Figure 3-4.) Added to it should be a safety factor of 10 percent, roughly another millimeter. This addition is necessary in order to prevent the stop from leaking light into the objective, thus weakening the darkfield effect. Actually, I cannot confirm the reliability of this widely recommended formula for determining the stop's diameter. In practice, not even a 12 mm stop produced a satisfactory darkfield image with the 6× objective. It is correct that there must be a minimum stop size. However, the stop size can be considerably larger than the backlens measurements suggest.

I use a dime, a nickel, and a quarter for all darkfield observations and photography with objectives giving 3.5× to 25× magnifications. If the coins are mounted with Krazy Glue (which does not affect the Plexiglass), the micronaut has a permanent set of darkfield stops. If the coins are used unmounted, much care has to be taken to center the coins each time they

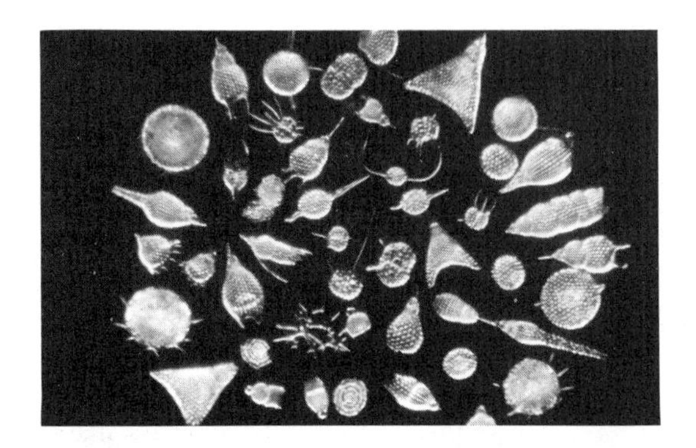

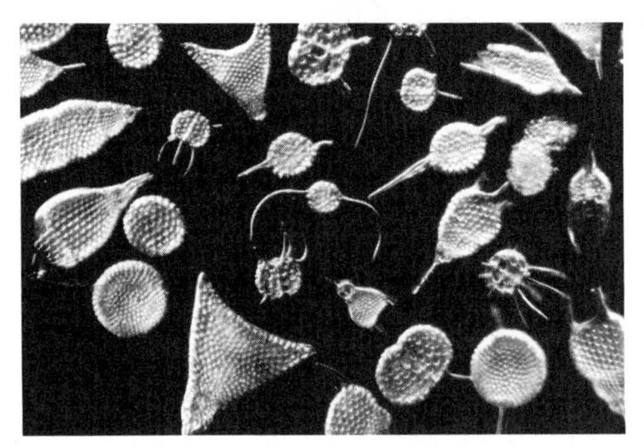

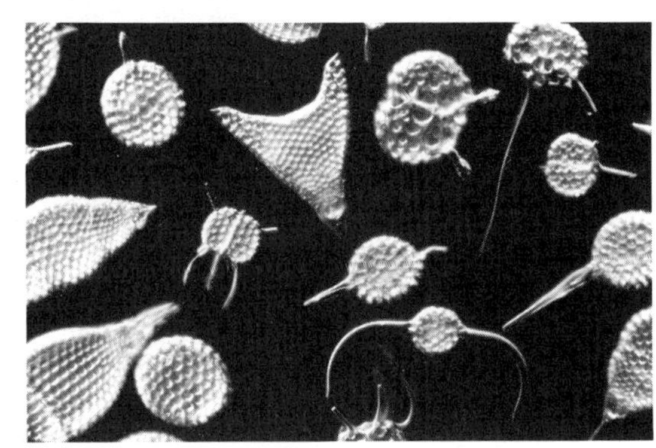

FIGURES 3-5, 3-6, and 3-7. Photomicrographs of *Radiolaria* with 3.5 X, 6 X, and 10 X objectives, respectively. Magnifications are 20 X, 32 X, and 44 X.

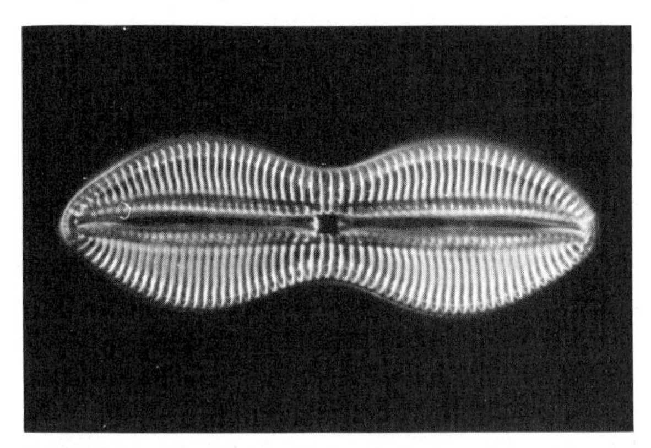

FIGURE 3-8. The diatom *Diploneis crabo* taken with a 25 X objective and darkfield illumination, 400 X.

are used. Their position can be checked by placing a mirror underneath the condenser, in the same way that the aperture iris is viewed when Köhler illumination is adjusted (see Figure 2-17).

Figures 3-5 through 3-7 illustrate some results obtained using coins as stops. A dime was used for Figure 3-5 (50×); a nickel was used for Figures 3-6 (90×) and Figure 3-7 (120×). The background of these photomicrographs is intensely black, and the *Radiolaria* are shown in good detail. For the 25× objective, a quarter provides a proper stop, as shown in Figure 3-8, a photograph of the diatom, *Diploneis crabo* (400×).

For objectives higher than 25×, condensers especially designed for darkfield are required. The photomicrograph of *Paramecium* (see Figure 3–1b) was taken with such a condenser. As compared with the photograph taken by brightfield illumination (see Figure 3-1a), Figure 3-1b provides a clearer look at the food vacuoles and even the yeast cells inside these vacuoles with which the animal was fed. The two dark spots near the lower edge of the cell are the contractile vacuoles in systole. At this point of the cycle the excess water has just been discharged.

Although the homemade darkfield is restricted to low power work, its potential is not exhausted with the described manipulations. Darkfield allows modifications that considerably extend its usefulness. The basic idea of the darkfield principle is to block light from entering the objective. So far, we have considered the effect of light entering the whole peripheral ring of the condenser. We can go a step further and restrict some, or even most of the peripheral rays. This leads to another way of "playing with the light," the *oblique illumination*. This method is also available to the naturalist through simple means.

Oblique Illumination

The idea of oblique illumination originated with a special condenser, invented by Ernst Abbe. In this condenser, the aperture diaphragm could be moved sideways in a controlled manner by means of a rack-and-pinion mechanism, directing the incoming rays to illuminate the specimen at an angle. Abbe's substage apparatus is no longer made. However, it is not too

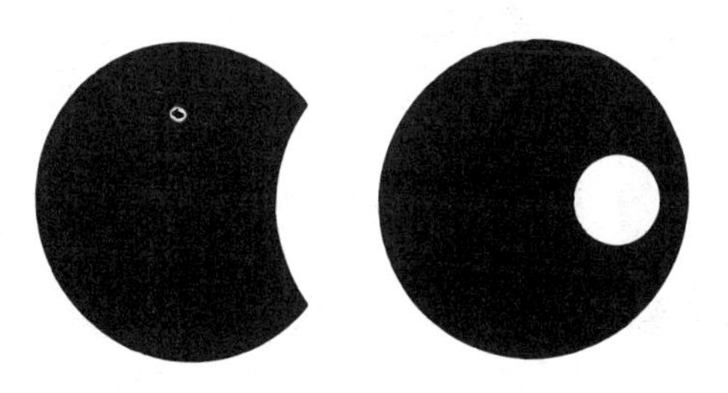

FIGURE 3-9. Two substage discs for oblique illumination.

FIGURE 3-10. Triangle cut-out for oblique illumination after Dieter Gerlach. (*Courtesy* Dr. Dieter Gerlach, Universitaet Erlangen, German Federal Republic.)

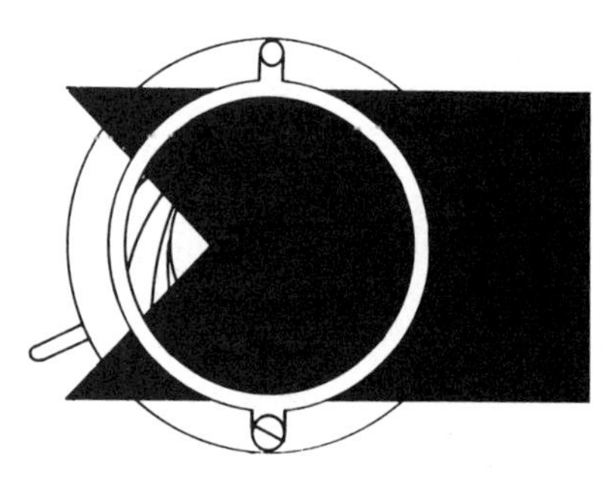

FIGURE 3-11. Salt from a salt shaker photographed with a substage "hole diaphragm," 35 X.

FIGURE 3-12. Salt from a salt shaker photographed with a triangle cut-out, light background, 35 X.

FIGURE 3-13. Salt from a salt shaker photographed with a triangle cut-out, dark background, 35 X.

FIGURE 3-14. Salt from a salt shaker photographed with a decentralized mirror, 35 X.

difficult to make a substitute by cutting an overexposed photographic sheet film to fit into the filter carrier, then perforating it with a punch or notching the edge to leave a half-moon opening through which the light can enter (see Figure 3-9). Rotate the opening to find the most favorable position for the specimen's illumination. The size of the half-moon or the hole depends on the objective to be used and must be determined by trial and error. Swinging in and out the top lens, lowering or lifting the condenser, slightly closing the aperture iris and/or the lamp diaphragm are all operations that affect the image and should be tried. A photomicrograph taken with such a "hole diaphragm" is shown in Figure 3-11.

Another gadget that allows one to vary the size and position of the little "window" that admits the light is a piece of cardboard with a triangle cut out at one end (see Figure 3-10). This piece of cardboard is inserted into the space between the filter holder and the condenser, and positioned in such a way that only a small passage for the light remains. By opening or closing the aperture iris, additional control of the image is possible. The width of the cardboard must be sufficient to cut off all light except that coming through the slot. The width should also be sufficient to prevent the cardboard from moving sideways. The peak of the triangle must be exactly in the middle between the edges of the cardboard. With this device, oblique illumination can be set up with a light background (see Figure 3-12) or, if the cardboard is pushed in a trifle more, with a dark background (see Figure 3-13).

Parenthetically, it should be mentioned that oblique lighting can also be achieved without any stops, holes, half-moons, or triangles, simply by bringing the mirror out of center. This method, however, is the least recommendable. While all ways of achieving sidelighting result in some unevenness of the image, sidelighting is most pronounced with mirror manipulations (see Figure 3-14).

Rheinberg Illumination

Having outlined the adaptation of the microscope for darkfield and other related methods of illumination, there remains one more variation: Rheinberg differential color illumination.

Most microscopical specimens are colorless. Chlorophyll-bearing algae are an exception. Some protozoa have a brownish tint, but otherwise the microcosm is a grayish world. In the 1830s, color was introduced, not so much for the sake of coloring as to differentiate structures in cells or tissues. It was found that certain dyes have the property of staining specific parts of a cell, such as the macronucleus in a protozoan. It was Christian Gottfried Ehrenberg (1795–1876) who started this technique which, during the last

150 years, has developed into a separate science of histology in which more and more sophisticated methods of chemical treatment have been devised for cell research.

Although staining with chemicals is an important technique, it offers few possibilities for the naturalist studying live objects. Limited applications are possible with so-called vital stains that stain, but do not kill the cell. These stains will be discussed in Chapter 4, "Your Specimen." Fortunately, the versatility of the microscope provides a means of bringing color into the colorless by *optical staining.*

The first method for this kind of coloring was suggested by the Englishman, Julius Rheinberg, in 1896. This method is essentially a modification of darkfield illumination. Instead of making the central stop dark and leaving the peripheral ring bright, Rheinberg replaced these with stops and rings of different colors. The effect is a color contrast between the specimen and its background. On Plate 2, the diatom *Arachnoidiscus ehrenbergi* appears yellow in the color of the outer ring, but appears on a green background in the color of the stop. A variety of color combinations suggest themselves. The central stop can be left black, but the outer ring can be half blue, half yellow. If preferred, the stop can be blue and the ring half green, half red, or the center can be left black and the ring divided into even three different colors (Plates 3 through 5).

Rheinberg filters are not available commercially, but it is possible to make the filter combinations out of colored cellophane. Three products lend themselves to this purpose.

1. Kodak Wratten gelatine filters come in 3 × 3-inch sizes and in a great variety of colors. Any large photographic supply store carries them.
2. Tecnifax Diazochrome film is a colorless material that can be colored by exposing it in daylight to ammonia vapors in a closed jar. During this procedure, it is necessary to avoid dipping the film into the ammonia liquid. A practical way of doing this is to place a piece of plastic sponge soaked with ammonia in the jar. Household ammonia from the supermarket or grocer is satisfactory. The film is left in the jar for about ten minutes or until it has reached a deep color. No washing or drying is required. The only negative aspect of this useful material is its cost. Diazochrome film comes in 8 × 10-inch sheets, twenty-five sheets to a package.

At least four colors would be needed. The trouble is that only two or three sheets will be required for ample experimenting. It would be a boon for a micronaut if the manufacturer of this material could make an assortment of a variety of colors available in one package. Such an assortment of colors (red, orange, yellow, green, blue, and purple) could be bought by a group of micronauts sharing the expense.

Diazochrome film is sold by Tecnifax Corporation, Holyoke, MA 01042, and by Aresco, Inc., 341 West John Street, Hicksville, N.Y. 11802.

Another source of colored cellophane in many shades is Belden Communications, Inc., 534 West 25th Street, New York, NY 10001, the U.S. distributor of the English "Lee Filters." They come in large sheets of 21 × 24 inches, and here again it is the minimum quantity available that causes problems.

Many products are packaged in colored cellophane wrappings. If saved, these wrappings may serve as inexpensive Rheinberg discs, provided the material is not too flaccid. Whatever its source, the cellophane or film has to be cut to size with a pair of manicure scissors. For the outer dimensions, the Plexiglass type of support can serve as a template; for the diameter of the central stop, a coin of the appropriate size can be used. If the stop is also to be colored, the inner edge of the peripheral ring can also be shaped with a coin to guide the scissors. When attaching the different filter sections to the Plexiglass disc with glue, it is important to prevent any light leaks. The central stop often needs a double layer of cellophane or film to achieve the Rheinberg effect. The difficulty of making a colored central stop can be circumvented by leaving that stop black, using instead two peripheral colors, each covering half of the disc.

Light Polarization

Another even more important method of optical staining is the use of polarized light. Polarization is based on the wave theory of light. It assumes that light propagates in waves that vibrate in all directions. If, however, such a ray passes through certain materials, the vibration is affected in such a way that the waves pulsate only in one plane; that is, they become plane-polarized. One such material is calcite and another is a chemical, iodo-sulphate of quinone.

The polarizing effect of calcite was discovered by William Nicol in 1828, who first applied its usage to the microscope. Naturally, Nicol prisms were expensive. They had to be cut into flat rhombohedron-shaped crystals. No naturalist could have afforded a polariscope. But in 1935, a college student by the name of Edwin Land took up work with the chemical iodo-sulphate of quinone, the polarizing property of which had originally been discovered by W. B. Herapath in 1852. Herapath's findings had no practical application. Each of the "Herapathite" crystals that he grew polarized singly, but not in a volume because they were not uniformly arranged. It was Edwin Land's achievement to have found a way of organizing these minute crystals in such a way that they could act like a single crystal. Moreover, he was able to embed the particles in a durable transparent material that could be easily shaped to the needs of a host of optical instruments.

This revolutionary achievement gave the contemporary naturalist the unique opportunity to apply polarization to his or her microscope. The application could not be simpler: all that is needed is a small piece of Polaroid material, which is available in 3 × 3-inch sizes from any large photographic supply store for just a few dollars. Out of this sheet, the one-piece polarizer is cut into a disc to fit the filter carrier of the condenser; another piece, the analyzer, is cut to fit in or over the eyepiece.

Light polarization is based on a basic principle. After passing through the polarizer, the plane-polarized beam, now vibrating in one direction (say north-south), goes through the condenser and, assuming that no specimen is yet on the stage, through the objective until it encounters the analyzer. Because the analyzer is made of the same material as the polarizer, it has the same property of polarizing any ray entering it to vibrate in one plane. If its direction of vibration is parallel to that of the polarizer, the beam will go right through, and the field will be bright. However, if the analyzer is rotated, the light intensity gradually decreases until, when the analyzer disc is at 90° in reference to the position of the polarizer, no light can pass through, and the field of view will be dark. This position is referred to as "extinction" or the "crossed" position.

With crossed "polars," some materials remain dark throughout an entire rotation of 360°. They are isotropic. Sodium chloride (NaCl), for instance, is isotropic. If such a crystal is put between crossed polars, it remains invisible, and the field of view remains dark. The reason for this effect is that the polarized light goes through the grain of salt unaffected, being stopped by the analyzer. Anisotropic substances, on the contrary, do change the light coming from the polarizer. Substances from this group, including some crystals, wood, starch, hair, and textile fibers, have the property of birefringence. In this property, light passing through such matter is split, and the two rays emanate at different velocities, one ray being retarded with reference to the other. When passing through the analyzer, these rays are brought back into one plane. However, due to the phase shift that they suffer, the fast and slow rays interfere, leading to the production of so-called interference colors.

White light consists of different colors of different wavelengths. The effect of the interference phenomenon is that several components of the white light are partially suppressed. In the emergent beam, these components are no longer in the right proportion to give white light; some colors are missing. As a result, the specimen appears in the colors of the remaining wavelengths. Color contrast results, which is of great value in identifying the material under observation.

A refinement of polarization is possible by adding so-called compensators between the polarizer and the analyzer. Various minerals, such as gypsum, mica, quartz, or selenite, often are used as compensators. Modern research microscopes provide a space where these plates can be inserted below the analyzer. However, a compensator can also be applied on the stage, directly underneath the specimen. Years ago, such plates were made by English manufacturers, but are not available any more. However, progress in plastic chemistry has provided the microscopist with a convenient substitute, transparent cellophane, which can be used if the purpose is strictly to achieve optical staining, without any scientific aim. Brilliant colors can be achieved using transparent cellophane.

Between crossed polars, our friend the *Paramecium* is mostly dark on a dark background (see Figure 3-1c). This means that its cell contents—the cytoplasm, nuclei, food vacuoles, contractile vacuoles, granules, and cilia—are all isotropic, not birefringent. However, there are particles that are brightly visible. These are inclusions, calcium oxalate crystals, a product of the animal's metabolism, and are a very significant detail not revealed by either the brightfield or darkfield methods (compare with Figures 3-1a and 3-1b).

Truly spectacular are photomicrographs of anisotropic objects such as the vitamin shown in Figures 3-15 and 3-16 and the piece of granite from Central Park in New York, shown on Plates 8a and 8b. As a thin section cut to 30/1000th of a millimeter, even granite becomes transparent. When it is photographed with brightfield, almost nothing can be seen (see Plate 8a). But if the same field is photographed between crossed polars, a contrast is created that helps the trained crystallographer or geologist to identify the kind of material he or she examines under the microscope (see Figure 3-16 and Plate 8b).

FIGURE 3-15. Vitamin B_1 crystals photographed with brightfield (50 X).

FIGURE 3-16. The same field photographed with polarized light (50 X).

Incident Illumination

Incident lighting is used to examine the surface of an object by means of the light reflected from it. For this type of lighting, an ordinary microscope lamp, raised or placed on books, is aimed at the object at an angle of about 45° or less if a special three-dimensional effect is intended. This illumination is not different from ordinary photography. Only low power lenses can be used because objectives with short working distances obstruct the light on its way to the object. Even with low power objectives, the specimen may throw too much of a shadow if placed on a light background. In this case, a second lamp from the opposite side is needed to minimize the shadow effect. It may be expedient to use a weaker illuminator as the second light source or to place it farther away from the microscope.

In setting up the illumination, the bulb's filament has to be focused on the specimen with the lamp diaphragm closed. For this purpose, the ground glass is removed temporarily. After focusing (the filament coils must show on the specimen), the ground glass is replaced and the lamp diaphragm is reopened to a certain extent. All of these operations are prompted by the intent of achieving the best possible image without excessive glare and contrast.

What may prove disturbing in incident light are reflections, which often interfere with the image. If this occurs, it is necessary to manipulate the lamp or lamps to locate the origin of the reflections. Sometimes a reflection comes from the very objective in use; then covering the front of the lens with black tape will help. As a precautionary measure, it is advisable to remove all objectives from the nosepiece except the one used. Another preventive is to place a piece of black paper or cardboard underneath the

FIGURE 3-17. A foraminiferan photographed with reflected light, 10 X.

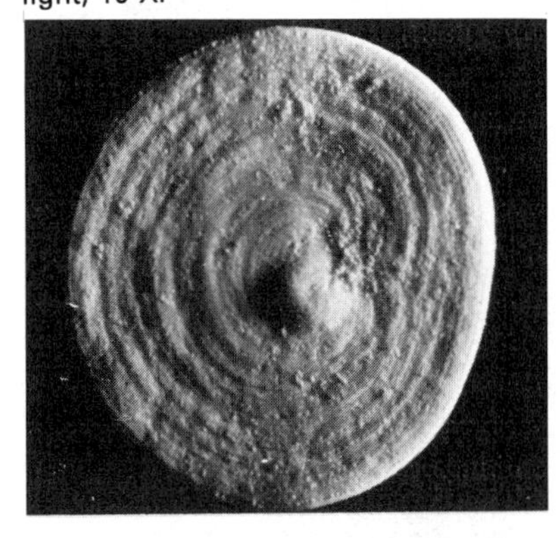

FIGURE 3-18. The same foraminiferan photographed with a combination of reflected and transmitted light, 10 X.

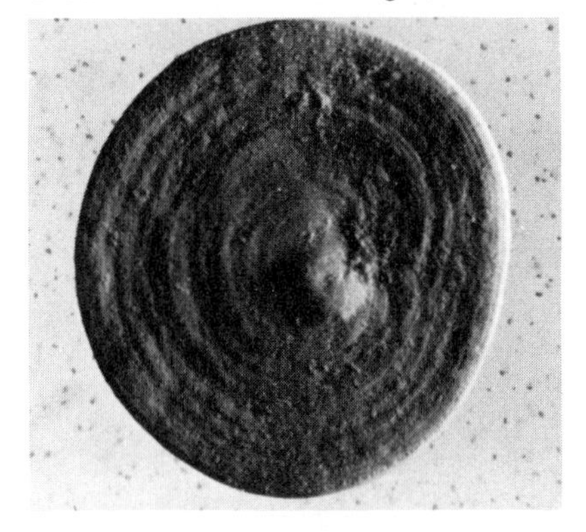

stage where reflections from the condenser may interfere. Also, turn out all overhead lights in the room.

An incident image can sometimes be much improved when weak transmitted light is combined with reflected light. The transmitted light can be modulated with color or neutral density filters. Figure 3-17, a photomicrograph of a *Foraminiferan,* was taken with reflected light. In Figure 3-18, the same specimen was photographed with combined light.

Modulation Contrast

A more recent advance in microscopical optics is the so-called Modulation Contrast. Developed by Dr. Robert Hoffman, a physicist, its merit consists of providing a three-dimensional image comparable to that obtained with oblique illumination. Its important difference is that it avoids the unevenness of oblique lighting that is so difficult to control, as Figure 3-14, and even Figure 3-11 indicate.

FIGURE 3-19. Diagram to show the principle of modulation contrast. (*Courtesy* Dr. Robert Hoffman.)

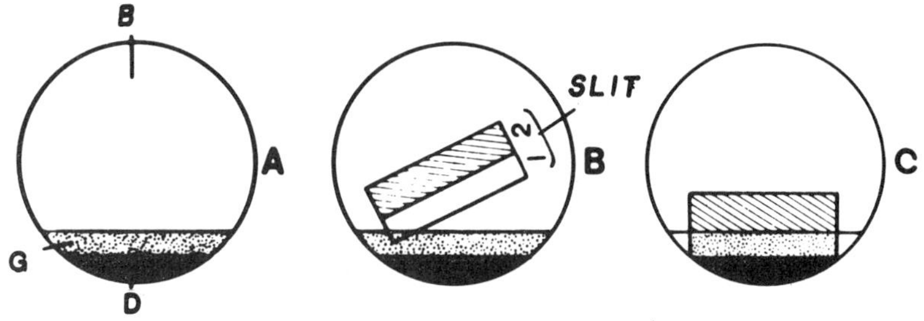

The three-dimensional effect is achieved by inserting a special filter into an ordinary brightfield objective. This plate, the modulator, has three segments of different densities: the first opaque, the second gray, and the third bright. In terms of density, the segments have respective transparencies of 0, 15%, and 100%. Underneath the condenser, a rectangular slit is placed that is made partly of Polaroid material, partly bright. The slit can be rotated and moved sideways. The idea is to coordinate the modulator and the slit, as shown in part C of the diagram in Figure 3-19. In addition, a polarizer is put over the light source or, if an old-fashioned microscope is used, into the lamp's filter holder. The polarizer functions in conjunction with the Polaroid part of the slit and serves to control the light intensity and/or the image contrast. It can also be rotated.

Any brightfield objective can be converted for modulation and still be used for brightfield illumination. The conversion of a microscope can be done in the inventor's factory: Modulation Optics, Inc., 100 Forest Drive at East Hill, Greenvale, NY 11548. The conversion is not inexpensive.

However, two papers have been published that suggest do-it-yourself solutions.[1]

An example of the image that modulation contrast can produce is presented in Figure 3-1d. The natural appearance of a freely swimming *Paramecium*, which gave the animal the popular name of slipperanimalcule, is well recorded. In the other photographs of *Paramecium* in this chapter, the animals, though living, are slightly flattened because they were photographed in order to disclose morphological detail, which can best be seen under a cover glass.

Phase Contrast

Living cells such as bacteria, yeast, starch, and some protozoa, are often so transparent that they can hardly be seen with brightfield illumination. In the early days of microscopy, the only makeshift to overcome this deficiency was either to kill and stain the organisms or to reduce the aperture diaphragm. As was pointed out in Chapter 2, the latter method is just the kind of adjustment that is taboo: by introducing diffraction effects, the resolution is ruined.

Phase contrast, the ground-breaking invention of Frits Zernicke, a Dutch physicist, for which he received the Nobel Prize in 1932, surmounts the difficulty. His solution is based on certain physical laws: If light passes through an unstained living cell, such as a protozoan, it suffers little or no loss of its intensity, but is affected in a different way. The organelles of a cell have different refractive indices. The macronucleus of a protozoan, for instance, has a higher index than the drop of pond water in which it swims about. These refractive indices have a specific effect on the light rays that pass through a substance: They suffer a retardation, the extent of which depends on the refractive index of the area they pass through. The higher the index, the greater the retardation. The latter is expressed in a phase shift of the light waves. In other words, the crests and valleys of the light waves are displaced against one another in a manner similar to the displacement that occurs when polarized light passes through birefringent material. The displacement or retardation by a living cell may amount to one fourth or one half of the light's wavelength, but our eyes cannot perceive these changes. Zernicke's accomplishment consists of having found a way to intensify these phase differences and translate them into amplitude or brightness differences.

Phase contrast requires special objectives and condensers. The objective must have a phase ring, a ring of magnesium fluoride that has the function of retarding the light rays passing through it. The phase ring is located in the back focal plane of the lens (see Figure 3-20). Below the condenser,

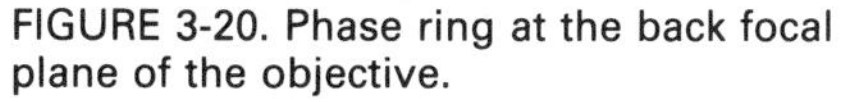

FIGURE 3-20. Phase ring at the back focal plane of the objective.

FIGURE 3-21. Annular ring diaphragm below the condenser.

a ring diaphragm is placed (see Figure 3-21). For setting up phase contrast, the phase ring and the ring diaphragm must be coordinated. This coordination is done by inspecting the backlens of the objective with a special telescopic eyepiece that enlarges the backlens. Both rings must exactly overlap.

Phase contrast is a most useful system. Because of its expense, I have devised a substitute that requires the acquisition of a 40× phase contrast objective only, while the ring diaphragm and the telescopic eyepiece can be made by a microscopist who has access to photographic darkroom facilities. With a 40× lens, a wide range of magnifications from 160× with a 4× ocular to 640× with a 16× ocular is possible. The telescopic eyepiece also comes in handy for other occasions when the need arises to view the backlens for darkfield illumination. The paper detailing this method was published under the title, "A Substitute Phase Contrast Attachment" in the *Transactions of the American Microscopical Society,* Vol. 96 (3), 1977, pp. 60–64.

Figure 3-1e was taken with a 40× phase objective by Carl Zeiss. Compare it with Figure 3-1a. The contractile and food vacuoles have much better contrast; the macronucleus (underneath the prominent contractile vacuole at the upper right) is also more distinct.

Interference Microscopes

Important advances have been achieved in the development of interference microscopes. These advances are actually refined phase contrast systems, with the difference in that the retardation is made possible with means other than the Zernicke system.

The basic idea of an interference microscope is to split the light into two bundles. One bundle is directed through the specimen, where it is subjected to a phase shift. The second bundle does not pass through the object and does not experience retardation. When the two bundles are joined again in the optical system, they interfere, resulting in a phase contrast which, if the light is polarized, can also be expressed in color differences.

There are two systems, both developed by Carl Zeiss. The *differential interference system after Nomarski* employs two prisms to split and rejoin polarized light. It provides a black-and-white image as well as color contrast. It is an important method, widely used today because it provides a three-dimensional image. As to the color rendering, it must be said that the color does not significantly add to the information. Therefore, only an example of its black-and-white potential is given here (see Figure 3-1f).

Another system is applied in the *interference microscope after Jamin-Lebedeff,* in which a birefringent calcite plate splits the light. One bundle of light passes through the specimen, while the other one goes through an empty area of the preparation. The bundles are again combined by a second calcite plate. Because one beam is retarded and the other one is not, they interfere when recombined. This time, however, the different refractive indices of the cell organelles are expressed in different colors. This makes the Jamin-Lebedeff interference microscope one of the most remarkable instruments of optical staining. It cannot be applied to tissues, however, because a secondary (ghost) image is inherent in this system, and the many secondary images produced by a continuous sheet of cells destroy the sharpness of the image. Nevertheless, when applied to single cells, this microscope gives significant contrast.

On Plate 1, a *Paramecium* is again shown, with the various organelles defined in color. The natural shape of the animalcule is preserved, since no chemical treatment distorted the cell. In this photograph we can distinguish:

the cytostome (mouth opening)	brown
the cytoplasm	yellow
the food vacuoles	green
ingested yeast cells	light green
the contractile vacuoles	white
the cilia at the edge	yellow
the bases of the cilia	brown dots
the macronucleus	brown
the micronucleus (rarely seen in the living animal)	purple

Indeed, a microscopical "cat-scan"—in color!

Fluorescence

Many rocks illuminated with "black light" glow, often in spectacular colors. The light, as pointed out before, is called black because its wavelength is so short that our eyes cannot perceive it. Being outside the visible spectrum on the violet end, it is also called ultra-violet light, or UV for short.

The phenomenon of glow, known as fluorescence, is also applicable to the microscope. Not only rocks, but many biological substances are sensitive to UV radiation. If a fluorescing matter is exposed to light containing rays of a wavelength below 385 nanometers (or 3850 angstroms) it emits light of a longer wavelength—blue, yellow, green, or red. This reaction is called *primary fluorescence.* A number of living things do have this quality, among them *Paramecium*, a fact not generally known except to specialists. The photomicrograph (see Plate 7) was taken with the light of a high-pressure mercury vapor lamp rich in UV. Ordinary household bulbs even of 200 watts or more are useless because they provide very little UV radiation. In the photograph, the cytoplasm has a weak green color; it is mostly the food vacuoles that are emitting rays of longer wavelengths, such as yellow and red. Taking such a photograph has its problems. Since electronic flash cannot be used, the animal must remain motionless for the duration of the exposure; that is at least a minute. During this period, the colors often fade. Shortly after the photographic session, the animal dies due to the deadly influence of the radiation.

Primary fluorescence has its limitations. These limitations led to the development of fluorescing stains, which opened up the wide field of *secondary fluorescence.* These stains are referred to as fluorochromes. They are selective, adhering to specific structures in the cell. It is not the structure itself that glows, but the absorbed dye. If the same *Paramecium* species as that shown on Plate 7 were subjected to a weak solution of Acridine Orange (1:2000, that is 1 gram dissolved in 2000 ml of water) it will fluoresce in far more brilliant colors (see Plate 6). This technique makes the following cell components visible: the macronucleus and the food vacuoles appear in green; the cytoplasm is clearly granular throughout and stains orange; a number of cell inclusions fluoresce yellow. These are the so-called neutral red globules, so named because they normally stain only with Neutral Red, a common biological stain. The function of the neutral red globules in *Paramecium* has not been identified yet, although they are assumed to participate in the digestive process of the animal.

The optical set-up for fluorescence requires the following accessories:

1. A proper *light source*. Only a high pressure mercury vapor lamp or a halogen bulb is suitable. The former is not only expensive, but also has a short laboratory lifespan of only about 150 hours; after this period, it may explode and damage its housing. Therefore, the mercury vapor lamp should be replaced before its maximum lifespan is reached. The halogen is less expensive, but its UV output is much smaller.

2. The light emitted by these light sources is white light, containing all wavelengths, including UV, and therefore must first be screened. A special filter, the *exciter filter,* serves this purpose. It excludes all wavelengths except the blue and the ultra-violet. The exciters for blue excitation are designated by the codes BG3, BG12, and BG38. Blue filters are used because the reradiated light is nearest to violet and offers the potential of producing all colors of longer wavelength: green, yellow, orange, and red. A green exciter can also be used (and is used if the purpose requires it), but the color variety is accordingly reduced. A green exciter can produce yellow and red, but not blue fluorescence.

3. After its passage through the exciter, the filtered light next passes through the specimen if transmitted light is used. In an incident set-up, the light is reflected and contains the colors emitted by the specimen, including unabsorbed blue and UV. Before the light is allowed to reach the observer's eye, however, another screening is necessary, which is performed by the *barrier filter.* Its purpose is to bar any remnants of the ultra-violet radiation, but to let the fluorescent light waves pass. This is important, because unfiltered UV would damage the observer's eyes. For the same reason, neither of the two bulbs should ever be looked at directly while in operation.

On a reduced level, it is possible to adapt a less elaborate microscope for fluorescence observation or photography. Again, the microscope must have a mirror to accomplish the conversion. The main problem is the light source, but the suggestion of Dieter Gerlach offers a solution.[2] He proposes the use of a slide projector equipped with a halogen lamp. In this case, the microscope would have to be elevated to the same level as the projector.

A projector is not a microscope lamp and would not permit either Köhler or Nelson illumination to be carried out. However, these illumination methods would be unnecessary. The foremost consideration and purpose is to get as much filtered light, with as much UV as possible, into the specimen. For this reason, the aperture diaphragm has to be kept open for fluorescence.

Only one exciter, the BG3 filter, is necessary for the beginning. Gerlach suggests a piece of cobalt glass as a substitute. This solution may be less expensive, but it is also more difficult to get. The BG3 filter, according to my experience, gives excellent results. Plates 6 and 7 were both taken with it.

For the barrier filter, numbers 50 or 52 are recommended. Barrier filters supplied by Zeiss have a diameter of 18 mm and can be dropped into the eyepiece to rest inside on the ocular's diaphragm.

With these relatively simple means, the experimental possibilities are numerous. Relevant objects for examination are bacteria, fungi, algae, protozoa, waterfleas, and others. Among the fluorochrome stains, Acridine Orange, Rhodamine B, Phloxinrhodamin, and Auramine O can be tried. An excellent introductory pamphlet has been published by Zeiss. This pamphlet, "Worthwhile Facts About Fluorescence Microscopy," written by H. M. Holz, is available for $2.00 from Carl Zeiss, One Zeiss Drive, Thornwood, NY 10594.

4
Your Specimen

In microscopy, it is not enough to master the microscope. No less important is the art of handling the objects to be observed. In fact, such skill is not only an art, but a science—the science of microtechnique, a huge and complex field. Its application requires glassware, chemicals, and gadgets of all kinds. But within the scope of this book and the needs of a home laboratory, we are restricted to simple procedures. Only temporary mounts will be discussed. Permanent mounts of biological material would require chemicals not generally available, such as fixatives of various kinds, stains and alcohol of different strengths. Temporary mounts, on the other hand, need a minimum of equipment, the most useful being:

Microslides (size 1–3 inches)
Round cover glasses (#1½ inches)
Rectangular cover glasses (size 24×40 mm, #1 thickness)
Small pipettes (medicine droppers)
Large pipettes (11 inches)
Needles for teasing, spoons, toothpicks
Scalpels and single-edged razor blades
Metric chemical balance
Plankton net (with hand-held stick)
Insect-killing jar with cotton packing, for saturating with alcohol
Screw-top jars and bottles of different sizes
Culture dishes

Graduated cylinders (to hold 100 ml and 200 ml)
Centigrade thermometer
Alcohol lamp
Lens paper
Biological vital stains
Distilled water
Spring water
Xylol or toluol
Balsam or Permount
Cedar oil
Turpentine
pH paper, as HYDRION pH Paper Strips pH 4–9

Not all of these items have to be acquired at once. Other necessary items will be mentioned as we go along.

STAINING

Some specimens do not need staining, especially if optical staining (polarized light or Rheinberg illumination) is available, or if nature provides the color, as in green algae. Other specimens are so transparent that little detail can be discerned with brightfield illumination. It is in such cases that phase contrast is desirable, although it is out of reach for the modest pocketbook. For such specimens, a few stains should be kept on hand. The most practical are Neutral Red, Gentian Violet, and Methylene Blue. The advantage of these stains is that they can be used also as *vital* or *intra vitam* stains, meaning that they color the organism (for instance a protozoan) without harming it, provided they are used in very diluted form. At higher concentrations, they will kill the animal, although they will stain it better.

CRYSTALS

Simple experiments can be performed with *crystals,* some of which can be found right in the kitchen. *Sugar* is one example. Dissolve some granulated sugar in a liqueur glass with hot water until no more can be dissolved, resulting in what is called a saturated solution. Put a drop or two of this solution on a slide and spread it out evenly. Then leave the slide in a dust-free place to dry, or recrystallize, which will take about two days. The resulting crystals will be colorless but, under polarized light (see Plate 5), they will show colors of many shades. The solution need not be saturated;

it can also be diluted. In this case, however, a different pattern of crystals results.

Another striking example is *salt,* sodium chloride. A saturated solution yields single cubes, as shown in Chapter 3, Figures 3-11 through 3-14. When a diluted, rather than a saturated salt solution is used, the salt crystallizes in a different way, although the basic crystal system remains the same. *Tears* are, with the effective help of a sliced onion, an easily accessible source of a diluted salt solution. They consist of only 1.98% dry matter and 98.2% water. The dry matter is mainly "kitchen salt," NaCl.

Tears, as Figure 4-1 shows, crystallize in an arboresque form. Although the form looks quite different from salt in the salt shaker, the basic molecular structure is the same. Sodium chloride belongs to the cubic system of crystals. This is clearly evident in tear crystals, in which the branches of the "tree" are strictly arranged at 90° relative to the direction in which the crystals grow when forming.

The shape of crystals also depends on the speed of crystallization. The process can be hastened by placing the slide on a warm plate or on a radiator in winter. But slow crystallization, as a rule, yields better crystals.

Many other water-soluble chemicals can be handled the same way: washing soda, borax, aspirin, citric acid, epsom salt, vitamins, and chemicals for photographic processing, to name only a few. The best way to find out is to try yourself. I once put a drop of an old, neglected eyedrop prescription on a slide, just out of curiosity. It provided me with one of the most photogenic crystals I have ever seen, looking just like chrysanthemums (see Figure 4-2). The material turned out to be Metaphen, an antiseptic.

FIGURE 4-1. Crystal from a tear, 20×.

FIGURE 4-2. Crystallized Metaphen, an eye drop prescription, 15×.

FIGURE 4-3. Condensate from tobacco smoke, 150×.

Another method of growing crystals is to melt the material over an alcohol lamp, then let it cool and recrystallize. A wax candle is a good object for such an experiment. Place a very small amount of wax on a slide, cover it with a cover glass, and then, holding it with a pair of pliers over an alcohol lamp, *gradually* melt the wax by intermittently withdrawing the slide from the flame. As the wax melts, it spreads under the cover glass; when the material has spread to a thin layer, withdraw the slide from the flame and let it cool. Very small crystals will cover the field of view.

Interesting results also can be obtained by holding a slide over the vapors or smoke of burning matter. When the gas touches the slide, it solidifies in a process called sublimation. Cigarette or cigar smoke are an interesting material to try. Just let a cigarette burn on an ashtray at an angle and arrange a microslide, for instance on a small laboratory tripod, in such a way that the smoke touches the slide before dissipating; then inspect the resulting yellowish deposit through the microscope (see Figure 4-3).

Starch grains are obtained easily from potatoes. Scrape a raw potato over a handkerchief, press out the juice, put a drop of the juice on a slide, and cover with a cover glass. There will be thousands of single starch grains of different sizes (See Figure 4-4). Then examine the specimen with polarized light. Each grain has a characteristic "cross," which is visible only between crossed polars.

The *onion* is a gratifying subject to observe as well. Peel off the outer dry skin, place a piece of the skin flat on a slide, and cover. When viewed, the mount will reveal numerous crystals scattered throughout the tissue (See Figure 4-5). These crystals consist of calcium oxalate, a product of the plant's metabolism. They appear in many colors, but again only under

FIGURE 4-4. Starch grains from a raw potato, 165×.

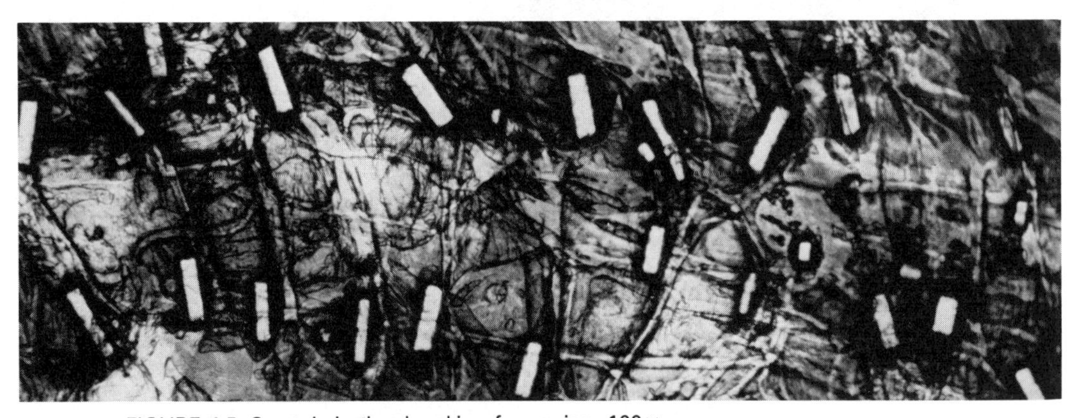

FIGURE 4-5. Crystals in the dry skin of an onion, 100×.

crossed polars (compare with Figure 3-1c, a micrograph of *Paramecium* taken with polarized light). Such crystals occur both in plant and animal tissues.

Another interesting subject is the thin transparent skin that separates the succulent layers of the onion bulb. This skin peels off easily from a quartered bulb and should be mounted in water to delay its drying out. The single layer of cells in this membrane allows one to study the single intact cell without letting it pass through a microtome, an instrument that cuts tissue into thin sections for microscopical examination. In the onion's membrane, the basic components of a plant cell—the nucleus, cytoplasm, and the outline of each cell—all are clearly visible (see Figure 4-6).

While on the subject of cell membranes, let us not forget the *leaf epidermis,* especially that of the lower surface of leaves. In lilies, for example, these layers are easily stripped off. Place one layer flat on a slide, then add a drop of water and a cover glass. When viewed through a microscope you will see the stomata surrounded by so-called guard cells. The stomata are minute pores that regulate the water content of a plant. Stomata close in response to environmental conditions to conserve water, if necessary, or open to permit evaporation of excess water through transpiration (see Figure 4-7).

Textiles are also easy to prepare for viewing. They can be viewed with or without polarized light. A nylon stocking provides a pretty pattern; so do linen and silk. Cotton wool is best prepared by teasing the fibers with two needles. All textile specimens are better covered with a second slide instead of a thin cover glass to flatten the fabric. Other tissues such as lens paper (an indispensable item for cleaning the optical components of the microscope) are also worth looking at.

Insects offer a wide variety of material for observation. The killing is best done with an insect-killing jar. These jars are available in two types, those that include cyanide, and those that include alcohol as the toxic agent. The latter is preferable. A vial cemented to the screw top of the jar is filled with cotton which, before use, has to be filled with ethyl acetate.

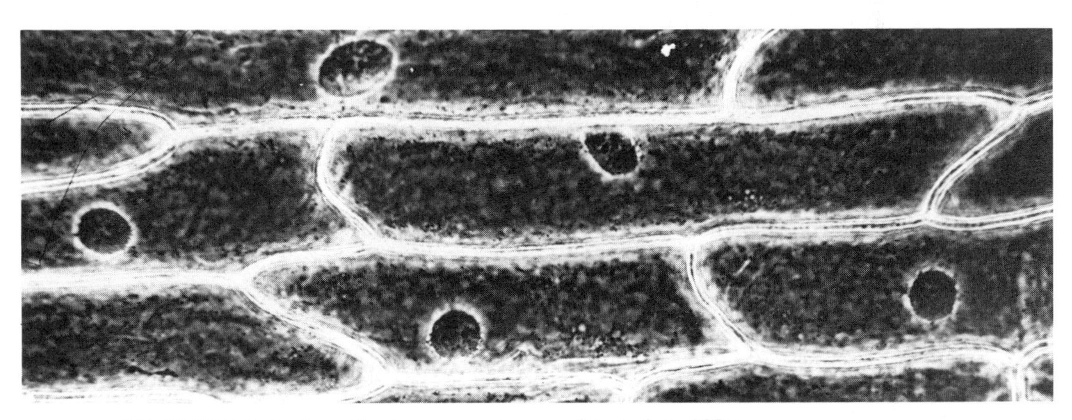

FIGURE 4-6. Cells from the inner membrane of an onion, 285×.

FIGURE 4-7. Stomata from the epidermis of a lily, 265×.

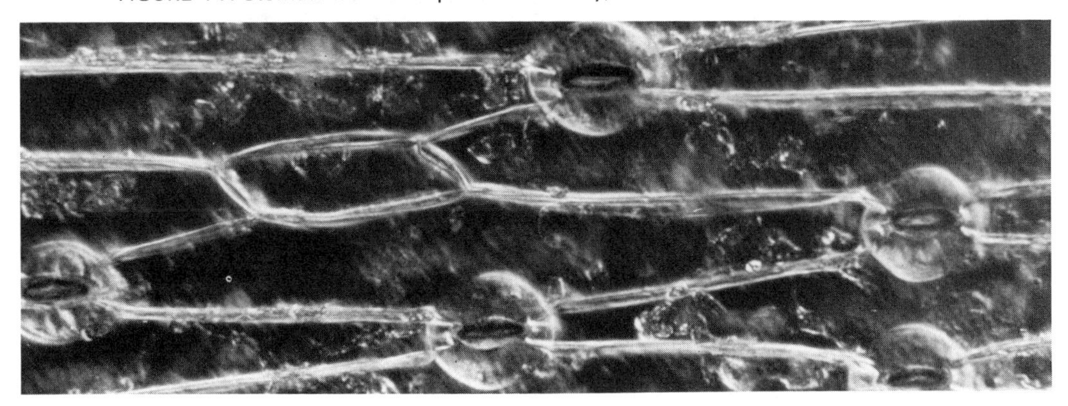

Very small insects like aphids, which can be found in almost any garden, can be made into slides as whole mounts without much processing. From the killing jar, transfer the aphids to turpentine. After an hour or so, put a specimen on a slide and let the turpentine evaporate. Now apply a drop of balsam, and position the insect with a needle. The needle must be moistened with xylol or cedar oil if it is to be dipped into the balsam during positioning, or the balsam will stick to the needle and impede your work. In order to prevent a delicate insect from being crushed by the cover glass, props should be used. Small bits of broken cover glass put beneath each corner of the cover glass are the most practical for this.

Many insect parts are so opaque that they cannot be viewed by transmitted light. In such cases, reflected light must be used, unless the special technique of potashing is used. This technique can be accomplished by applying potassium or sodium hydroxide, both caustic alkalis that destroy all internal organs of the insect and leave only the bleached exoskeleton. Yet a number of insect parts do not require such rigorous treatment. Wings from flies, mosquitoes, bees, wasps, and many other insects can be directly mounted in balsam without preliminary preparation, and thus incorporated into a collection of permanent slides. The same holds true for the scales from butterfly wings. Just hold a wing over a slide and "dust" it off with a soft brush. A drop of balsam and a cover glass finish the job.

Especially fascinating are the compound eyes of insects, with their thousands of six-sided facets. Some dragonflies have up to 28,000 of these facets in each eye; houseflies have 4,000 in each eye. An easy way to attract flies is to put a piece of rotting fish into a small jar. When a few flies have entered to feed, hold a killing bottle over the collection jar. In trying to escape, the insects will fly into the killing bottle, which then must be closed quickly. It is to be hoped that the flies will have left some eggs behind, providing another interesting specimen from which maggots may hatch a few days later.

The eyes of the fly can be observed *in situ* by placing the fly, legs down, on a slide, and adjusting the lamp for reflected light. An eye also can be amputated with a razor blade and put, surface down, under the microscope for viewing with transmitted light. In this case, however, not much is visible, because the inner content of the compound eye interferes with the observation. Here nature comes to our rescue—instead of a fly eye, we can use a crab eye (see Figure 4-8). Crabs, lobsters, and other crustaceans have the same kinds of eyes as insects, yet on a much bigger scale, so that they are much easier to handle. At the time of year when crabs are harvested, the fish market should be able to provide you with one or two crabs that are already dead, though well enough preserved to be useful for the microscopist. Snip off the eye stalk with a pair of scissors, cut off the facets with a razor blade, then hold the eye by its edge with a sharp tweezer (size #5). Place the eye in a 0.4 percent solution of NaOH (sodium hydroxide) and

FIGURE 4-8. Detail of a crab's eye, 265×.

FIGURE 4-9. Image of Anthony van Leewenhoek photographed through a crab's eye, 535×.

carefully brush away the soft organic matter inside the cup. An eye-held magnifier such as watchmakers use will be necessary during this process unless a stereomicroscope is available. What remains is the lens element of the compound eye. When placed in a drop of water under the microscope, it is possible to view the facets from the inside at much higher magnification than with reflected light.

Such a mount, though only a temporary one, can also be used for an interesting photographic experiment. An obvious question is: How good are the lenses of an insect's or a crab's compound eyes? Very good, I must say. I once photographed a small transparency of Anthony van Leeuwenhoek that I arranged underneath the condenser. Each facet lens produced a separate, quite sharp image of this great microscopist, as shown in Figure 4-9.

What specimens can we use from *our own bodies*? One of your hairs is an obvious suggestion. And don't forget your dog's or cat's hair. These specimens can be handled easily in a way similar to that used in preparing

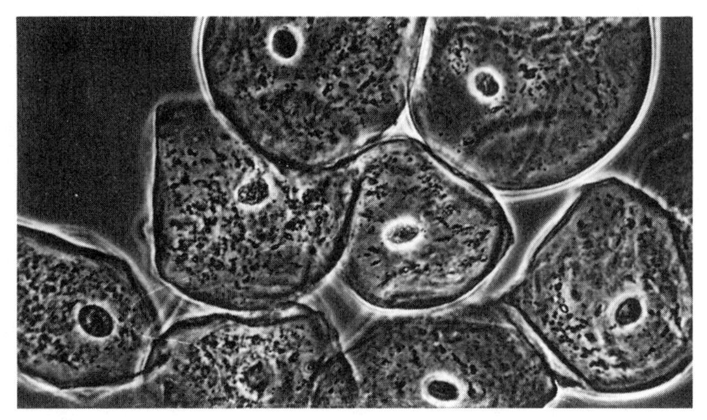

FIGURE 4-10. Cheek cells, 570×.

the direct balsam mounts of fly wings. Hair, though, should be dipped into alcohol, and then xylol, to remove any dust or fat that may adhere to it.

Easily accessible *cells* from our own bodies are the "cheek cells" (see Figure 4-10), the epithelium cells that form the mucous membrane inside our cheeks. We can recover them by scraping the area with a wooden tongue depressor (available in any drugstore). These cells are transparent, and must be stained. For a temporary preparation, mix the matter with a little saliva on a glass slide, and spread it. Let it dry, then pass it quickly over an alcohol lamp three times, a procedure that kills and fixes the cells. Apply Gentian Violet to the slide for one minute, rinse gently in tap water, and dry. Gentian Violet is sold as a 1 percent solution under the brand name "Viogen" in drugstores. It must, however, be diluted to prevent overstaining; one drop in 10 cc of water makes a suitable solution.

A similar procedure can be followed with the matter that adheres to our teeth, even if we brush them religiously twice or three times a day. This is the material that Leeuwenhoek investigated, thereby becoming the first man to see and describe bacteria.

For viewing these bacteria in living condition, remove some of the matter from one of your molars with a toothpick, spread it with saliva on a slide, cover, and cut down the light by closing the aperture diaphragm. With a 40× objective and a 10× or 15× eyepiece, long rods will be seen moving about in a leisurely waving motion. These are *Bacilli subtilis,* the giant bacteria that live in everybody's mouth without doing any harm. The staining of these bacteria is done the same way as for cheek cells.

Hay also yields a variety of bacterial organisms. Put a handful of hay in a half-gallon jar of water and let it stand. The film on the surface of the water that forms after a few days is a dense accumulation of bacteria.

The *air,* too, is a rich source of bacteria and such microorganisms as fungi. However, to capture them, culture media must be bought or made.

A simple formula (after Corrington) calls for the addition of 10 grams of Bacto-Peptone, 5 grams of meat extract, and 5 grams of sodium chloride to 1000 cc of spring water, all of which should be heated until the ingredients have dissolved. Then add water to the 1000 cc mark to replace that lost by evaporation, and sterilize the mixture according to the method later described. The resulting preparation is called nutrient broth, a liquid medium in which single cells can be observed after inoculation from agar plates.

A different medium, nutrient agar, is prepared on plates. Cook 15 grams of powdered agar in 1000 cc of spring water until it is dissolved, which takes 30 to 40 minutes. While it is boiling, add and stir in the ingredients previously mentioned for nutrient broth. Again, add water to compensate for evaporation. Cool the mixture to 60°C, then add the whites of two eggs.

For sterilizing, put mixture into two bottles and close the bottles. If the bottles are closed with screw caps, there is danger that they may burst during the subsequent boiling. Therefore, screw caps must not be used. Instead, use a cotton plug capped with a piece of gauze tied around the bottleneck with a piece of string. Then boil the bottles in a pressure cooker at 15 pounds of pressure for 20 minutes.

After sterilization, the agar is poured into Petri dishes and cooled. Then the dishes can be left standing open for a while, held briefly against the exhaust of a vacuum cleaner, or put into an open window, or sprinkled with dust from the floor. You can also sit with the open plate for about 20 minutes in the subway, a crowded railway station, or bus terminal. After the plates have been "contaminated," they must be covered and left alone for a few days. Then colonies of different kinds of bacteria and fungi will appear. If one of the colonies is transferred into the nutrient broth by means of an inoculating loop, its cells will multiply rapidly and can then be viewed as smears under the microscope. In bacteriological work, everything that touches the cells—all tools, pipettes, and glassware—must be sterile.

Though many bacteria grow at room temperature, the cultures, whether in nutrient broth or on agar plates, grow faster in an incubator. Such a piece of equipment can be constructed, for example, from old microscope cases occasionally obtainable from second-hand microscope dealers. Two holes must be drilled into the top of the case, one to accommodate an electric bulb, the other one a Centigrade thermometer. The strength of the bulb has to be determined by trial. Depending on the size of the box, a bulb of between 15 and 60 watts will be required. The ideal condition would be to keep the temperature at a steady 37°C. Sophisticated micronauts may be able to connect a thermostat to the incubator.

The exposed plates may also contain some fungi colonies, but if you are after fungi, cornmeal agar from biological supply companies is a better medium. Instructions for its use will be found on the bottle.

The most fascinating subjects for microscopical observation are the living invisible fauna and flora that can be found in almost any water accumulation—ponds, lakes, ditches, swamps, puddles, indoor aquaria—and even in water from flower vases in which flowers have been kept for some time.

I usually set out on my collection trips with two half-gallon screw-top jars and fill them three-quarters full with material from different sources as I go. With a large spoon or a plankton net on a stick, I skim the surface of various water accumulations, especially if green scum or filamentous algae are present, which may entrap waterfleas, copepods, larvae, and protozoa. I also take up some mud, rotten leaves, and debris. Mud is important in wintertime. This is where resting cysts can be collected. Once home, food and warmth are added as, for example, one gram of white flour or pressed yeast, emulsified in a liqueur glass. As a rule, these cysts hatch. Inexpensive Brewer's Yeast tablets (three tablets per liter), crushed and sprinkled on the surface of the water are also effective.

Another useful material for the microscopist to culture is *horse manure*. Basically, this is chewed, half-digested grass or hay, enriched by the digestive juices of the animal. Not more than one crushed horse dung ball is added to a previously prepared hay infusion concocted by boiling 2 grams of hay in 1 liter of water, even tap water. Such cultures have provided me with such delicacies as *Diloptus, Euplotes, Stylonychia,* and other Protista commonly encountered in pond collections. Here again, we owe the harvest to the presence of cysts. In our motorized times, horse manure has become a rare item for the city dweller. As a New Yorker, I usually get the material from riding academies or at Fifth Avenue and Central Park South, where horses line up and wait to take tourists through Central Park.

Micronauts near the seashore have an additional source for collecting microorganisms: the *ocean*. There are two possibilities: to tow a plankton net through the water or to dig specimens out of the sand. The latter mode may even be the better and easier one. There are very interesting organisms to be found in the interstitial spaces between the sand grains. The only time for collecting is at ebb tide. When the water recedes, the animals that normally live in this environment become accessible; others, like marine diatoms, radiolaria (in warmer waters), and protozoa are trapped in the sand. Dig a hole in the ground with a small shovel until water accumulates in it. Then collect the water in a jar and take it home. Marine organisms are more difficult to keep alive outside of their natural environment; therefore, a quick examination is advised. Here is an area where new discoveries are possible, especially on lonely beaches in far-away lands.

Freshwater collections are also shortlived, even if fed. To keep them going, subcultures have to be established. Part of the organisms present have to be transferred into a fresh medium. For this purpose, it is not necessary to obtain pond water from the original source, but simply to prepare an artificial medium in which the animals are likely to thrive. The same procedure is required if cultures are bought from biological supply houses. The quantity received will be small, and is meant to be used as a seed for a larger harvest. For this purpose, an artificial medium has to be prepared in which the animals (or plants) can be expected to reproduce.

Recipes for artificial media are as numerous as species of protozoa, whose tastes and requirements are very different. The literature abounds with media formulas, sometimes very complex ones.[1] Fortunately, there are a few simple media applicable to a large number of protozoa. They are based on hay, lettuce leaf, or turnip infusions, with or without the addition of flour, wheat grains, rice grains, oat flakes, yeast, fresh milk, powdered milk, bread, cacao, or other items we humans also appreciate as part of our diet. Edgar P. Jones,[2] who investigated the dietary needs of *Paramecia,* mentions these foodstuffs as useful supplements to revive declining cultures. I would add Brewer's Yeast, an inexpensive vitamin available in drugstores. It comes in tablets that must be crushed and sprinkled over the culture. Use three tablets to a liter. In applying these edibles, we have to keep in mind that we are dealing with microscopic organisms. Overfeeding may be harmful. Too much food affects the pH. If the medium becomes too acid, the whole culture may die in a short time. Supplementary additions should not exceed one gram per liter. If cultures are kept in 200-ml dishes, only sprinkle small quantities.

These feeding recommendations are valid for a wide range of animalcules that are bacteria feeders, many of which are ciliates. If a specific species is to be cultured, the best way of finding a suitable culture method is to consult a textbook on protozoa, such as *Protozoology,* by R. Kudo, Springfield, IL: Charles C. Thomas, 4th ed., 1960. The index can be used to locate those researchers who have worked on the organisms. The paragraphs on "materials and methods" will probably provide the needed information, as well as the pH preferred by the animal in question.

My own method for culturing *Paramecium* uses "Protozoa Pellets." These pellets are the equivalent of dried lettuce leaves and are available from the Carolina Biological Supply Company, Burlington NC 27215.[3] I use three pellets, boil them in one liter of tap water for five minutes, then cool and filter the broth. Subsequently, I add about 3 to 4 grams of unboiled hay in order to let the bacteria develop in the medium. Two to three days later, depending on the temperature, the medium becomes slightly turbid, a sign that the bacterial population has increased. At this point, I measure the pH. The symbol, "pH," indicates the acidity or alkalinity of a solution. Distilled water is neither acid nor alkaline and has a pH of 7. Any value between seven and 0 is acidic; any value from 7 to 14 is alkaline. Measurement is done with strips of litmus paper, available, for instance, under the brand name "pHydrion Paper." When dipped into the solution to be measured, it turns red if the solution is acidic, light green at pH 7, deep green at pH 8, and blue at pH 9, if strongly alkaline. *Paramecium* grows best at pH 7 to 7.5. Using a pH strip to check the previously mentioned culture formula, it will be found that the medium is too acid. This must be corrected if one wishes to grow *Paramecium*. The correction is done by adding a 5% solution of sodium hydroxide, NaOH, which has to be prepared beforehand. It must be added cautiously, drop by drop with a pipette, and the pH must be checked after each addition with a fresh strip until one strip yields a light green color. Then the medium is ready for inoculation. About 200 *Paramecia* are sufficient to start a culture. Before inoculation, but after the pH has been adjusted, I remove the hay; otherwise, the medium has a tendency to become acid again. The pH of a going culture must also be checked from time to time.

This formula is just one prescription for culturing a number of ciliates. The Carolina Biological Supply Company adds to each shipment of invertebrates a little manual that suggests suitable culture methods. Among the 22 available kinds of ciliates, amoebas, and flagellates, there are many that thrive on a wheat medium. This medium is made by the addition of three or four previously boiled wheat grains to a culture volume of 200 ml. A hay-wheat medium is made up of two grains of wheat and two 3-cm stems of hay. Not less than 15 of the listed 22 protozoa species grow on these two media. As glassware for such cultures, Carolina sells special dishes that are 4½ inches in diameter and 1 3/4 inches deep, which can be stacked on top of each other.

Many protozoa are extremely active. Their activity makes their observation at medium and high magnification difficult. Not only size, but speed is magnified by the microscope. *Paramecium* is one of the fastest. With several hundred cilia that propel it through the water, the moment it crosses the field of view it is already gone.

There are, however, a few situations in which the animals are stopped by natural causes: if air bubbles are present, they become gathering places for the oxygen-hungry animals, which push one another to get closer to the bubble. The same happens when they find a detritus flake with a source of bacteria to feed on. Otherwise, they are always on the move.

Because air bubbles and detritus flakes are not at hand when they are needed, frustrated observers have devised ways to slow down single-celled animals (not only *Paramecia*) with mechanical or chemical means. If, for instance, two pipette drops of culture fluid are put under a 24×40 mm cover glass, the animals, at first, have ample room to swim about. As the fluid evaporates, they become more and more hampered. They also become flattened, which is often a desirable effect, because it is this condition in which the organelles are best revealed, especially if some of the new illumination techniques described in Chapter 3 are applied. The photomicrograph of *Paramecium* on the cover of this book, for example, was immobilized by this method.

Another mechanical mode of retarding the movement of one-celled animals consists of using a viscous matter, such as methyl cellulose. It is available under the name "Protoslo" from the Carolina Biological Supply Company (catalog number 88-5141). Protoslo comes in small bottles with a pointed mouthpiece through which a small quantity can be squirted out to make a ring on a microscope slide. A drop of medium containing the specimens is then placed in the ring's center and covered. As the sticky, nontoxic matter spreads, the animals are restrained as the Protoslo mixes with the medium.

Another way of quieting one-celled animals is the use of narcotizing chemicals. Many of these chemicals are unobtainable by most individuals because they contain cocaine or chloroform—although in extreme dilution—and will therefore not be mentioned here. However, there is one readily available concoction with an effectiveness that will surprise no one: 3% methyl alcohol, or less. Yes, even ciliates and flagellates can get drunk!

5
Photomicrography

Photomicrography is the art of recording images produced by a microscope on photosensitive material. The creation of a good microscopical image is therefore a precondition for a good photomicrograph. The means of mastering the microscopial part of this task has been treated in the preceding chapters. The rules for setting up a microscope are the same whether they are followed for the purpose of observation or recording, although in the latter case they may have to be followed a bit more strictly. Any oversight of the basic principles, even if its result is not apparent to the eye, will show in the final print or transparency. The illumination must be even, and the optical elements of the microscope must be clean and free of dust particles.

The main difference between general nature photography and photography through the microscope is that there are so many variables in the conditions the microscopist faces. Assuming he or she has only the minimum four objectives, the 3.5×, 10×, 20×, and 40×, and only two eyepieces, the 6× and 10×, there are already eight different possible situations that require eight different exposures. When one adds the variables of the filter factor and the film speed to be used, the challenge becomes more evident.

The way to confront these problems is the subject of this chapter. However, basic darkroom techniques, such as film developing, black-and-white or color printing, and slide making have been ably and completely explained in the excellent PHalarope Book by Stan Osolinksi, *Nature Photography: A Guide to Better Outdoor Pictures* (Englewood Cliffs, N.J.: Prentice-Hall, 1983), and, therefore, will not be discussed here.

Photomicrography can be done on 4-×-5-inch sheet film or on 35-mm roll film. The large format is used primarily for research. For the naturalist, the roll film is, without doubt, the more convenient, more versatile, and less expensive choice, and will be the only size considered here. A single lens reflex camera is suitable for photomicrography. Because images at medium and high magnifications tend to be dim, it is advisable to have the common matte glass screen changed for a clear one. No camera lens is required for photography through the microscope. Actually, readers who already own a single lens reflex camera will be able to use its body only. A microscope adapter takes the place of the lens. To the best of my knowledge, all of the great names in single lens reflex cameras, such as Nikon, Leica, Olympus, Minolta, Canon, Contex, and perhaps others make such adapters available.

The simplest type of microscope adapter is a tube that can be attached with one end to the camera body in the same way that a lens is connected, while the other end can be fastened to the microscope tube with the eyepiece in place. Other adapters are more elaborate, having a beam splitter inside and a viewer. Looking into the adapter, the outline of a screen indicates the size of the field of view that will be covered by the camera. An important feature of this viewer is a focusing device. This device consists of fine cross hairs in the center of the screen. Its purpose is to adjust the viewer for the observer's eye. If the cross hair lines (sometimes small, narrow double lines) are sharp and the image of the specimen, focused with the fine adjustment of the microscope, is in focus, the photograph will be sharp too. The focusing of the viewer has to be done only once for the microscope's user or users. The focusing knob is numbered −6 to +4 (on the Nikon), and each user must remember the setting or make a note of it according to his or her eyesight. The focusing knob should also be checked each time the microscope is used for photography because it has a tendency to get out of line.

Microscope adapters are usually marked with numbers, such as 1/3×, 1/2×, or 1×. These numbers indicate that the adapter contains a lens element, a correction lens, that reduces the magnification provided by the objective and the ocular by one third, one half, or not at all. A photomicrograph taken with a 40× objective, a 10× eyepiece, and a 1/2× adapter gives a magnification of 200× on the negative. If the picture is enlarged two times in printing, the final print will show the original microscopical magnification of the object. The amount of reduction is called the camera factor.

The adapter is a convenient device. The observer can sit down to look through the microscope instead of standing to peer through the camera finder, which is tiring, especially if living specimens are being watched. The adapter serves as a kind of substitute for an expensive binocular.

An advanced type of microscope adapter has a built-in, automatic exposure device. This is the *dernier cri* of microscopical technology. Such an

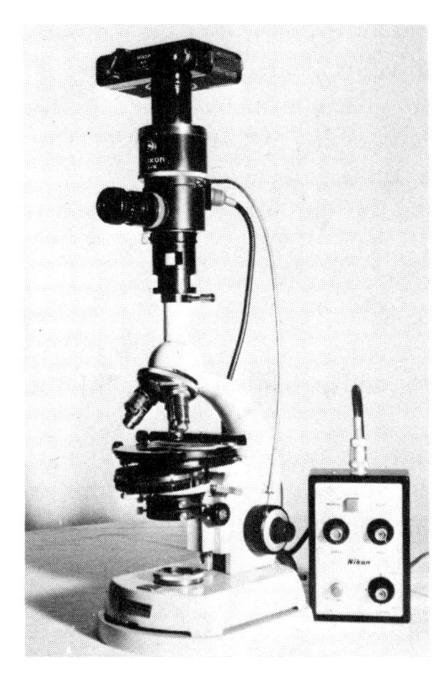

FIGURE 5-1. The author's latest photomicrographic set-up, a Zeiss Standard microscope and a Nikon microscope adapter, including an automatic exposure device.

automated microscope, my latest, is shown in Figure 5-1. All the photomicrographer has to do is to set the ISO,[1] the speed of the film being used on the control board, then press the cable release. The microscope in the picture is a Zeiss instrument; the automatic adapter, the Microflex AFM, is made by Nikon. When selecting the brand of the camera body, it is immaterial whether it has a built-in exposure meter or not. Exposure meters in single lens reflex cameras can be activated only with the lens in place.

An important subject in photomicrographical techniques is the use of filters. They are among the few tools available to improve the microscopical image by photographic means. Filters are put into the holder of the light source, into the carrier of the condenser, or simply placed over the light source of microscopes with built-in illumination.

White light, as mentioned before, consists of a whole range of different colors, each having a specific wavelength. These colors can be separated by dyed glass or gelatin filters. With a technique called spectrophotometry, it is possible to measure the range of colors a given filter allows to pass and which colors it absorbs. Figures 5-2, 5-3, 5-4, and 5-5 show such spectrographs.

The white areas indicate, on the abscissas, the wavelength transmitted and, on the ordinates, the intensity of the transmission. The black areas show the absorbed wavelengths. The four colors, purple, green, red, and blue are represented by their respective "Wratten"[2] numbers, 35, 61, 29, and 50. With the color shade, the graph changes slightly. A booklet issued by the Eastman KODAK Company, "KODAK Filters," 2nd edition, 1981, lists more than 120 spectrographs of different filters and filter combinations.

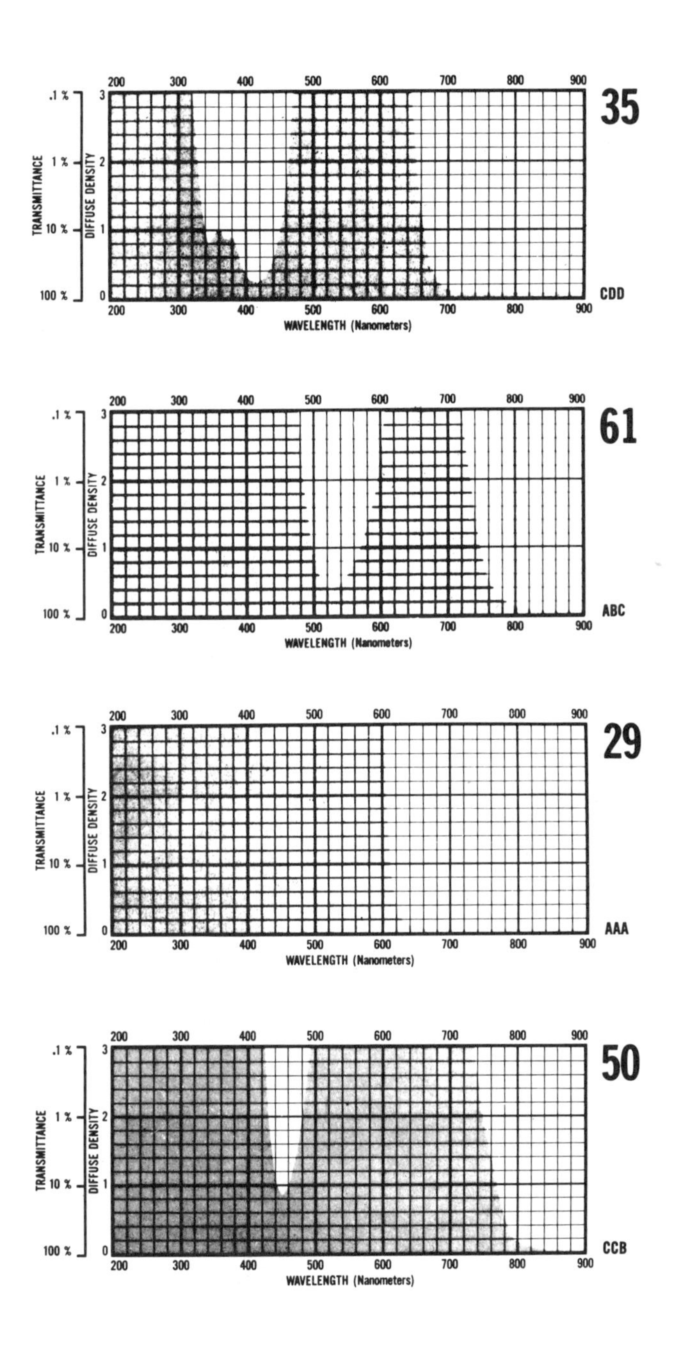

FIGURES 5-2 to 5-5. Spectrophotometric curves. (© Eastman KODAK Company. From KODAK Publication No. B-3, *KODAK Filters for Scientific and Technical Uses,* Second Edition, 1981: Spectrophotometric curves for Nos. 29, 35, 50, and 61. Used by permission.)

The performance of an optical filter can be measured in two ways: its light transmittance or its opacity or density. High transmittance corresponds to a low optical (here called diffuse) density and vice versa. While the transmittance is expressed in percent, the opacity is based on a mathematical formula that conveys the opacity value in numbers as the reciprocal of the transmission. The numbers on the ordinates mean that a density of 1 corresponds to a transmittance of 10 percent, one of 2 to 1 percent, one of 3 to 0.1 percent. This is done for convenience in mathematical computing.

These curves show a great deal of red transmission. This should not confuse us. Only light below 700nm wavelength can be perceived by our eyes *and* by a panchromatic film emulsion. The border of 700nm is the limit of the visible spectrum, as is the ultraviolet at its opposite end. Any higher wavelength is infrared radiation and of interest to those who do infrared photography.

Though this branch of photography plays an important role in medical photography, it is of limited application in photography through the microscope. It requires special film material. Rays of infrared radiation come to a different focus than the visual focus, a difficulty that is troublesome to overcome. Finally, as was pointed out in a previous chapter, the shorter the wavelength, the better the resolution.

What are the properties of the filters shown in the spectrophotometric graphs on page 71?

Figure 5-2 (#35) is a purple filter in the blue and violet range, with some red coming through.

Figure 5-3 (#61) is a green filter, one of the standard contrast filters for photomicrography.

Figure 5-4 (#29) is a red filter.

Figure 5-5 (#50) is a monochromat with a narrow wavelength band in the deep blue but also with a low transmission needing long exposures.

While each filter has its own Wratten number, groups of basic colors are sometimes distinguished by the following letters:

A	Orange-red	E	Orange
B	Green	F	Red
C	Blue-violet	G	Yellow
D	Purple	H	Blue-green

What are the practical applications of these filters?

1. The filters allow us to control the light intensity. If a filter transmits all wavelengths, but to a limited degree, it is used to cut out too much light. Such filters are called neutral density filters. Filters are available in a full set of ranges, from a density of 0.1 to 4.0. The most useful neutral density filters for photomicrography are those with densities of 0.3, 0.6, 0.9, and 1.2.

2. Filters can improve image contrast, their most important application in black-and-white photomicrography. No matter how well a microscope is adjusted, the result still depends on the quality of the preparation. Suppose we have a blood smear under the microscope that is weakly stained, showing the cells in a pale pink. If a black-and-white photograph is planned, a green filter of the group B should be put into the light path. If the stain is very weak, a dark green filter is called for; if the contrast requires only a slight improvement, a light green filter would be appropriate. The basic rule for contrast control is the use of a color *complementary* to the color of the preparation. Green is complementary to red, blue to orange, violet to yellow, and vice versa.

3. Filters can improve definition. If a very narrow part of the visible spectrum is transmitted, as in Figure 5-5, monochromatic light is produced which, as we have seen in Chapter 2, reduces chromatic aberration and thus improves resolution. Very little light goes through this filter. A photomicrograph taken under these conditions would need a very long exposure.

Definition can also be enhanced if a filter *corresponding* to the color of the preparation is used, provided the contrast is satisfactory.

These rules cannot be applied in color photography. The use of a green filter on a weakly stained blood smear would make the whole photograph green, including the background. The possibilities of manipulating color images for contrast or detail are few. Here the exposure becomes the critical factor. Too long an exposure results in a washed-out picture; too short an exposure results in too dark a picture. Light filtering in color photography has a different function: to adapt the light to the requirement of the film emulsion used.

Incandescent light sources, as well as daylight and electronic flash (but not light from fluorescent lamps) are distinguished by the concept of *color temperature*. This is a measure of the light's quality. The differences are expressed in "Kelvins," named after William T. Kelvin, (1824–1907), the British physicist who established this system. Kelvin values range from about 2600°K for an ordinary 100-watt electric household bulb, to 20,000°K for a blue sky. Microscope illuminators are also characterized by their color temperatures, as follows:

Photoflood lamp	3400°K
100-watt coil-filament lamp	3100°K
6-volt ribbon-filament lamp	3000°K
Quartz iodine halogen lamp	3200°K
Daylight and electronic flash	5500°K

The light-sensitive emulsion of color film gives correct color rendering only if it is exposed to light of the color temperature for which the film is made.

Since it is not feasible to produce different emulsions for every possible color temperature, the quality of the light has to be corrected. This correction is done with filters.

Kodachrome 25 and 64 film, for example, are made to give correct color with daylight or electronic flash. If these films, which require light of a color temperature of 5500°K, are exposed to light of a photoflood bulb of 3400°K, the result will be unacceptable. The photomicrographer has two possible choices in which to correct this situation: either to use a different film or a different color conversion filter. The proper film would be Kodachrome II Type A; the proper conversion filter would be an 80B. If a tungsten lamp of 3200°K is used, Ektachrome 50 or 160 film would be the right choice or, with daylight type film, a conversion filter of 80A. The problem is actually not as complicated as it may seem because the needed information is listed on the instruction leaflet included with each film roll. The following data are based on a table from the Kodak publication, *Photography Through the Microscope,* Seventh Edition, 1980:

TABLE 5-1
Kodak Filters for Kodak Color Films*

	Tungsten Type B Film, 3200 K	*Tungsten Type A Film 3400 K*	*Daylight Type Film 5500 K*
Light Source	*Filter No.*	*Filter No.*	*Filter No.*
6-Volt Ribbon-Filament (about 3000°K)	82A	82C	80A & 82A
6-Volt Coil-Filament (about 2900°K)	82B	82D	80A & 82B
100-Watt Coil-Filament (about 3100°K)	82	82B	80A & 82
300- to 750-Watt Coil-Filament (3200°K)	none	82A	80A
Tungsten Halogen (3200°K)	none	82A	80A
Photoflood Lamp (3400°K)	81A	none	80B

*©Eastman Kodak Company. From KODAK Publication No. P-2, *Photography Through the Microscope,* Seventh Edition, 1980, p. No. 50. KODAK Filters for KODAK COLOR FILMS.

The use of conversion filters, of course, lowers the speed of the film and requires an increase in the exposure. This information is also provided by the manufacturer with each roll of film. The following guide has been compiled from Kodak's instruction sheets.

TABLE 5-2
Decrease of Film Speed with Use of Conversion Filters

Kodak Film	*Color Temperature of Light Source Used in Kelvin*	*Conversion Filter Used*	*Film Speed ISO (ASA)*
Kodachrome 25 (Daylight)	3200°	80A	6
Kodachrome 25 (Daylight)	3400°	80B	8
Kodachrome 64 (Daylight)	3200°	80A	16
Kodachrome 64 (Daylight)	3400°	80B	20
Kodachrome 40, 5070 (Type A)	3200°	82A	32
Kodachrome 40, 5070 (Type A)	5500° (Daylight)	85	25
Ektachrome 50 (Tungsten)	3400°	81A	40
Ektachrome 50 (Tungsten)	5500° (Daylight)	85B	32
Ektachrome 100 (Daylight)	3200°	80A	25
Ektachrome 100 (Daylight)	3400°	80B	32
Ektachrome 160 (Tungsten)	3400°	81A	125
Ektachrome 160 (Tungsten)	5500° (Daylight)	85B	100
Ektachrome 200 (Daylight)	3200°	80A	50
Ektachrome 200 (Daylight)	3400°	80B	64
Ektachrome 400 (Daylight)	3200°	80A	100
Ektachrome 400 (Daylight)	3400°	80B	125

Microscopists must, of course, know the color temperature of their own illuminators. The most practical light source would be one of 3200°K. With this light source, there are two tungsten films (Ektachrome 50 and 160) that can be used without any conversion filter, and five daylight films (Kodachrome 25 and 64; Ektachrome 100, 200, and 400) that need only a single conversion filter, the 80A. Though the initial cost of a 3200°K lamp is higher than that of one with an ordinary coil filament, it pays in the long run and simplifies picture-taking.

One other type of filter for color photomicrography must be mentioned briefly: the color compensating (or CC) filters. They have been developed to correct what is called the reciprocity failure inherent in all light-sensitive emulsions. A film can be exposed for 1/8th of a second at f11 or for 1/4th of a second at f16;.in both cases, the negative receives the same amount of light. This is called the law of reciprocity. This law, however, is only valid within a certain range of short exposures. Color films are made so that exposures of fractions of a second give the best results. Yet it often happens in photography through the microscope that long exposures are needed. This is true for color as well as black-and-white material, though it is less pronounced with black-and-white. Since color film emulsions consist of three layers, each layer has a different chemical response to prolonged exposures. As a result, the color balance can be affected.

This reaction is called reciprocity failure. It can be corrected by using CC filters. These filters are available from the Eastman Kodak Company. They come in six colors—cyan, magenta, yellow, red, green, and blue—and in densities ranging from 0.5 to 50. A need for these filters is indicated whenever the transparency or color print has an allover light tint of yellow blue, or another unwanted shade. In case of a yellowish hue, for instance, a CCB filter is recommended (B stands for blue, the complementary color for yellow). The filter's density is indicated by numbers, such as CC10B or CC15B. To weaken an unwanted color, a CC filter of its complementary color is necessary. The need for CC correction is not a frequent problem with a good light source and/or a fast film. According to my experience, very short exposures with electronic flash do not affect the color balance markedly. A pamphlet, "E-1 Reciprocity Data: Kodak color Film," gives detailed information about the reciprocity problem. It can be obtained by your photographic dealer or ordered from the Eastman Kodak Company, Rochester, NY 14650, for ten cents.

Film emulsions for black-and-white photomicrography must be panchromatic because of the ability of filter use to control contrast and definition. The following 35-mm black-and-white films, all from Eastman Kodak, are available:

Panatomic-X	very fine grain	ISO 32
Plus-X pan	very fine grain	ISO 125
Tri-X pan	medium grain	ISO 400
Technical pan	very fine grain	ISO 125

The Technical pan film is a high-contrast film useful in photographing unstained specimens.

Considering the many variables faced in photomicrography, it is advisable to standardize the film generally used. I routinely use Plus-X pan. If the light level is low at high magnification, I switch to Tri-X pan, but only when photographing living material with phase contrast, interference contrast, or fluorescence.

Films for color photography have been mentioned in the preceding paragraphs on filters. They are distinguished by the light source for which they are made, their resolving power, and the speed that is indicated in their names. All come in rolls of 36 exposures, some also in short rolls of 20 exposures. The latter are convenient for test shots. Among the tungsten-balanced emulsions, only Ektachrome 160 is sold in rolls of twenty exposures. Of the daylight balanced films, Kodachrome 25 and 64, as well as Ektachrome 64, 200 and 400 are available in short rolls.

All color films mentioned are reversal films, yielding positive transparencies. However, prints can be made from them. The color transparencies can also be converted to black-and-white negatives from which black-and-white prints can be produced. Kodacolor is a negative film material with a speed of 80 ISO and balanced for daylight.

What about instant photomicrography, you may ask? It sounds tempting, especially for a newcomer to the microscope. There are two systems available, one developed by the Polaroid Corporation and the other by the Eastman Kodak Company. Both require sheet film cameras to produce black-and-white or color prints of $3\frac{1}{4} \times 4\frac{1}{4}$, 4×5, or even 8×10 inches. Polaroid also offers black-and-white film packs that deliver not only prints, but negatives. This is a convenient feature in that it provides the possibility of yielding an indefinite number of prints made by conventional photographic means. Incidentally, at this writing, the first press releases have appeared in the newspapers announcing technology for instant photography, including photomicrography, on 35-mm film rolls with Polaroid material. The user would have to obtain a special processor to auto-develop the rolls.

Can instant photomicrography be recommended to the neophyte in microscopy? Not wholeheartedly. First of all, the initial cost is considerable because a sheet film camera is necessary to operate the existing systems, at least until the 35-mm device has proved its merits. The film, a continuous expense, is also costly. The beginner will be better served instead by acquiring an exposure meter to overcome the greatest hurdle in photomicrography: the determination of the exposure time.

Instant photomicrography is not easier, but more difficult than traditional methods. The operative guidelines are less precise. Several exposures have to be tried until a satisfactory print is produced unless the operator has already had comprehensive experience in dealing with the countless situations encountered in microscopy.

Finally, in an effort to learn a new technology, it is not advisable to start with an advanced system that adds to the existing variables. In analyzing an unsatisfactory print, it would be more difficult to pinpoint the cause of the failure and whether it originated in the handling of the microscope or of the camera.

Finding the proper exposure is one of the more difficult problems in photomicrography. No formula exists that could help determine the proper exposure. In addition to the variables mentioned before (magnification, filter factor, and film speed), the following undefined conditions influence the exposure:

1. *The intensity of the light.* Even if the light intensity is constant, as with a photolamp (disregarding voltage changes in the line), it depends on the opening of the field diaphragm, and especially on the aperture diaphragm, each being determined by circumstances that cannot be precisely specified. For low-voltage lamps working on transformers where the operator sets the voltage level, it is advisable to eliminate one variable by standardizing the voltage to six or eight volts, although this is not always feasible.

2. *The density of the specimen* is another condition to consider. If the object is very dense, longer exposures are necessary in order to record any detail in it. The size of the object in relation to the size of the film frame must also be noted. For example, there are occasions in darkfield illumination when a small, brightly illuminated object has to be photographed on a black background. In such a case, even an automatic exposure meter will be misled. Such an exposure meter "sees" a primarily black area, and therefore gives the film a long exposure. An example of such a problem situation is shown in Figure 3-4, the photomicrograph of the backlens of a 6× objective under darkfield lighting. In this case, the image of the millimeter squares was actually very small, occupying not more than one centimeter of the frame, so that the exposure had to be shortened considerably. If the object is dark on a bright background, the meter sees a lot of light, leading to underexposure, so that the object will look like a silhouette. Costly automatic exposure meters, which measure selective spots on the frame, can take care of such problems. Without such an instrument, the photographer has to determine the best exposure by tests.

The only way to overcome these difficulties is to build up a record of exposures made under the different conditions you meet while exploring the invisible. As a beginner, you would have to start from scratch with a 3.5× objective, a 10× eyepiece, the available camera adapter, and a roll of Plus-X pan film, always making sure to set up the microscope with strict Köhler illumination. Trial exposures should be taken, and the following specifications for each exposure should be recorded on an index card or notebook:

Specimen
Objective
Eyepiece
Camera factor
Total magnification
Method of illumination
Filters
Film (ISO)
Developer
Light source
Exposure meter reading (if any)
Exposure in seconds or more
Comments

The very first trial exposure should be made with brightfield illumination and without a filter (except a density filter if the light is too intense). By eliminating the filter factor, a basic exposure value can be established. Filter factors are known and can be applied by increasing the exposure accordingly. The factors of the Kodak Wratten filters most frequently used for a panchromatic film such as Plux-X pan are:

filter A	29	red	factor 8
filter B	61	green	factor 8
filter C	47	blue	factor 10
filter XI	11	yellow-green	factor 3

For the test with a 3.5× objective, the following sequence of trial exposures is recommended: 1/100, 1/50, 1/15, 1/4, 1/2, and 1 second. The test strip should be developed according to the instructions of the manufacturer (Kodak), fixed, washed, dried, and inspected. A frame is well exposed if the white areas in the negative, which will appear dark in the enlargement, show some detail, unless, of course, these areas were black in the specimen. The two or three best exposures should be indicated on the file card for future reference. The file cards should be filed alphabetically according to the subject.

This test procedure must be repeated when higher magnifications are applied or if the method of illumination is changed. The following sequences (all with the 10× ocular) are suggested:

for the 10× objective: 1/50, 1/30, 1/15, 1/4 1/2, 1, 2 seconds
for the 20× objective: 1/15, 1/8, 1/4, 1/2, 1.2 seconds
for the 40× objective: 1/8, 1/4, 1/2, 1, 2, 3 seconds

If an exposure meter is used, measure the light intensity over the ocular. Although light meters are a great help, they have to be calibrated, because they indicate the *light intensity,* which then has to be translated into seconds or fractions of a second.

What Went Wrong?

This question is frequently asked when newcomers to the microworld scrutinize the results of their first efforts to record their discoveries. Assuming that the instrument and its accessories are in good order, eyebrows may be raised because of unexpected blemishes in the picture. Following are common shortcomings.

The Illumination Is Uneven

One side of the picture is darker than the other side, or the center is lighter or darker than the periphery, causing a disturbing vignetting effect. Both possibilities indicate imperfect Köhler or Nelson illumination. The sidewise irregularity may be caused by faulty centering of the bulb in the lamp housing, the mirror, and/or the condenser. If the condenser has no centering screws, the fault is located in the mirror and/or the lamp. The vignetting effect is caused if the condenser is improperly focused, either too low or too high.

The Picture Is Out of Focus

The sharpness of a photograph may suffer from vibrations, which may come either from the outside or from the shutter. Drawers in the table or bench on which the microscope rests should never be moved while an exposure is in progress.

In older microscopes, on which focusing is done by lifting or lowering the tube, the fine adjustment may get out of position if it does not withstand the weight of the camera body and the attached microscope adapter. In this respect, the design of modern microscopes, in which the stage and not the tube is moved, is a definite improvement. Advanced microscopical accessories for interference contrast or incident fluorescence would not support the additional weight of the attachments.

A photomicrograph cannot be sharp if the image of the specimen and the reticle crosslines, as seen in the side viewer, are not sharp at the same time. Special care must be devoted to the setting of the diopter ring with which the crosslines are focused. If in doubt, it is wise to check the focus through the camera finder.

Lack of sharpness can also be caused by glare, which results from the excessive illumination that occurs if either the aperture diaphragm or the field diaphragm is not properly closed. Glare is present if a haze covers the image.

Another condition that affects the sharpness is caused by inadvertently putting fingerprints on the optical components of the instrument. Fingerprints cannot be detected as such on a print, but they leave their mark in a veiled image. The same can be expected if an oil immersion lens is not kept clean. Such tainted optics should be cleaned with lens cleaner solutions. Toluol or xylol should not be applied because they may dissolve the cement that holds the lenses in place.

If fingerprints are not a frequent trouble, dust is. Dust shows on pictures as out-of-focus particles of odd-shaped configurations. These artifacts can spoil the best picture. Dust is most disturbing on specific glass surfaces of the optical sytem. A critical place is the ocular, especially at high magnification. An eyepiece has three surfaces where dust inevitably shows: the top surface of the eye lens, and both surfaces of the field lens. The same is true for the analyzer for polarized light if it is placed inside or on top of the ocular. Such dirt particles can be detected by rotating the eyepiece. Other critical places are the cover glass, the condenser, and the corrective lens or prism inside the microscope adapter.

Because the tube is sometimes left open without an eyepiece, dust may fall into and settle on the backlens of the objective. Therefore, it is a good practice to blow out objectives from time to time with a dustoff can of compressed air. Do not wipe the backlens of an objective with an alcohol- or xylol-wetted cotton swab. These solutions, though appearing clean, may leave a residue that is difficult to remove because of the inaccessibility of the lens.

Other places in the system to keep clean are the condenser, the mirror, and the filters. Artifacts on filters are easily traced by moving the filters back and forth while in the light path.

Now that the technical aspects of photomicrography have been covered, the duly informed micronaut can turn to the discoveries. The word *discovery* has a twofold meaning. Strictly, it means "to obtain, for the first time, sight or knowledge of anything existing already but not perceived or known" (*Webster's New Collegiate Dictionary,* 2nd edition, Springfield, MA: G&C Merriam Co., 1953, p. 237). With a few exceptions, this is not what the following pages are all about. To discover also means to learn, to find out what is unknown to oneself. It is in this sense that I set out, years ago, to discover for myself the world of the invisible. You will go places you never suspected to exist, which are inhabited by strange residents with strange habits. It is in this sense that the subsequent expeditions are to be understood.

part
two

6
"I Eat—Therefore I Am"

After a popular lecture on Descartes' philosophy, a listener raised his hand and asked, "Descartes said, 'Cogito ergo sum,' that is, 'I think—therefore I am.' Couldn't he as well have said: 'I eat—therefore I am'?"

Bent over the microscope watching the busy coming and going of the animalcules from my microzoo, I am often reminded of this incident. Indeed, the animalcules seem to have nothing else on their minds but their menus. The search for food dominates their very being. And nature's prolific imagination has found quite a number of ways and means to enable each species to get the nourishment it needs.

The second part of this book will be devoted mostly to the living components of the microworld. It will be filled with strange and unfamiliar names. Into which drawer of the vast collection of life on our planet do they belong?

The following table is a brief introductory survey that roughly classifies the subjects of our expedition, whether they are animal or plant, to which group they are related, whether they are unicellular or multicellular, and in which chapter their story is told:

Survey of Organisms Discussed in Part II with Reference to Their Taxonomic Positions in the Animal and Plant Kingdoms

Name	*Taxonomic Position*	*Uni* cellular	*Multi* cellular	*Animal*	*Plant*	*Chpt*
Algae	Chlorophyll-bearing plants not differentiated in roots stems and leaves	×	×		×	8

Name	*Taxonomic Position*	*Uni*	*Multi*	*Animal*	*Plant*	*Chpt*
		cellular				
Arachnoidiscus	diatom	×			×	3
Barbulanympha	flagellate	×		×		8
Blepharisma	ciliate	×		×		6&7
Bursaria	ciliate	×		×		6
Chlorella	green alga	×			×	8
Colpoda	ciliate	×		×		7
Convoluta	worm		×	×		8
Cryptocercus	insect		×	×		8
Cyclops	crustacean		×	×		7
Daphnia	crustacean		×	×		6
Didinium	ciliate	×		×		6&7
Dionea musc.	carnivorous plant		×		×	6
Diplodinium	ciliate	×		×		8
Diploneis	diatom	×			×	3
Foraminifera	amoeboid protozoa	×		×		3
Fungi	Chlorophyll-lacking plants, having, if microscopic, a filamentous body (Mycelium)	×	×		×	8
Hydra	a polyp or coelenterate		×	×		6
Intestinal Protozoa of Cryptocercus	flagellates	×		×		8
Isotricha	ciliate	×		×		8
Lacrymaria	ciliate	×		×		6
Lichens	composite plant consisting of a fungus and an alga		×		×	8
Paramecium	ciliate	×		×		
Pelomyxa	amoeba	×		×		
Pilobolus	fungus		×		×	8
Planarian	flatworm		×	×		7
Podocyrtis	radiolarian	×		×		3
Pyrsonympha	flagellate	×		×		8
Radiolaria	amoeboid protozoa	×		×		3
Saccinobaculus	flagellate	×		×		8
Saturnulus	radiolarian	×		×		8

Name	Taxonomic Position	Uni cellular	Multi cellular	Animal	Plant	Chpt
Spirochaetes	bacteria	×			×	8
Stentor	ciliate	×		×		9
Trichonympha	flagellate	×		×		8
Turbatrix	worm		×	×		7
Tyroglyphus	mite		×	×		8
Urinympha	flagellate	×		×		8
Utricularia	carnivorous aquatic plant		×		×	6

Chaos chaos Catching *Paramecium*

One of the most ravenous animals of the microworld is a large amoeba, which is known under various names: the giant amoeba, scientifically *Pelomyxa carolinensis,* and sometimes, because of its shapelessness, *Chaos chaos.* It was the Swedish naturalist, Carolus Linnaeus (1707–1778), who coined the last name, in frustration at the amoeba's peculiarity of constantly changing its shape. Linnaeus was the first scientist to put some order into the animal and plant kingdoms by developing a system of classification.

When food is scarce, *Chaos chaos* is rather motionless, but still alert, with its bulges, or pseudopods (meaning "false feet") at rest. Yet, as soon as a drop of water containing a good supply of *Paramecia* is added to a slide on which *chaos chaos* is being observed, the listless lump of protoplasm becomes visibly excited. Even before the *Paramecia* have a chance to reach the amoeba, life seems to return to the pseudopods. It is probably the commotion stirred up in the drop of water by the *Paramecia* that the amoeba, a very sensitive animal, is able to perceive.

Although a giant among protozoa, the amoeba has no definite size. When not extended, it may be no larger than 1 mm in diameter; stretched out, it can reach a length of 5 mm, thus becoming visible to the unaided eye. Still, the whole animal is only one cell—a huge one as cells go—which is capable of performing all the functions necessary to sustain its life and to reproduce. It has approximately 1000 minute nuclei, each 20 microns (20/1000 mm) large. The nuclei are too small to be visible in the illustrations.

When reproducing, the single cell divides into two or more individuals. It can divide without any nuclear changes; each new individual simply retains some of the nuclei distributed at random throughout the cytoplasm.

If you watch an active amoeba through the microscope, you will see that the cytoplasm constantly streams, as the blunt pseudopods slowly protrude here and there. The pseudopods serve the animal both as means of locomotion and as tools in capturing food. How the amoeba changes its shape—the mechanism used in the formation or withdrawal of pseudopods—is unknown.

The giant amoeba is one of the rarer organisms of the animal world. In general, to find it in nature is purely accidental. Its appearances are sporadic and have been reported from North Carolina, Virginia and Tennessee; even from a shallow pond in New York City, which, however, has since succumbed to progress. Fortunately, humanity, whose conscience is so heavily burdened with the extinction of many animal species, came to the rescue of this protozoan. It was taken to laboratories where culture media were developed on which the animal thrives, thus creating ideal living conditions for it in sterilized dishes all over the world. And why? Because this animal is exceptionally well suited for research. As a large protozoan, it is easy to work with and lends itself to the study of all kinds of problems of the free-living cell. The specimen pictured here (see Figures 6-1 through 6-7) was by no means collected from a pond; it came straight from a refrigerator of the Columbia University Anatomy Department.

The giant amoeba will devour other protozoa besides *Paramecia* if they are not too lively and vigilant to be caught. But *Paramecia* have a habit that makes it especially easy for them to be caught. When placed on a slide containing a giant amoeba, the *Paramecia* will first swim about aimlessly, but sooner or later one is bound to collide with the amoeba. And this is precisely the event the *Paramecium* depends on to find a potential source of bacteria that may serve it as food. Thus the *Paramecium* investigates the situation. Unsuspectingly, it probes the soft body of the amoeba. It is thigmotaxis, the orientation through contact and touch, that dooms the *Paramecium*. The result of the probing is that the amoeba's appetite is further stimulated. The streaming of the protoplasm in the pseudopods is accelerated.

FIGURE 6-1.

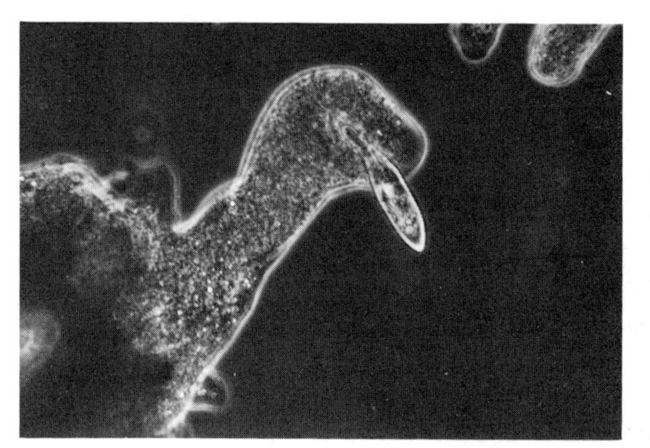

FIGURE 6-2.

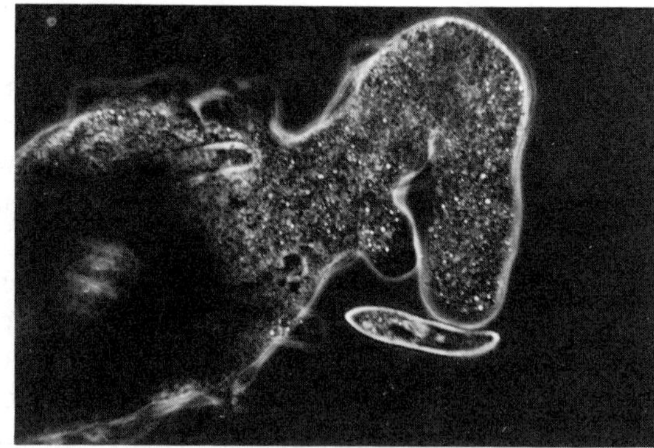

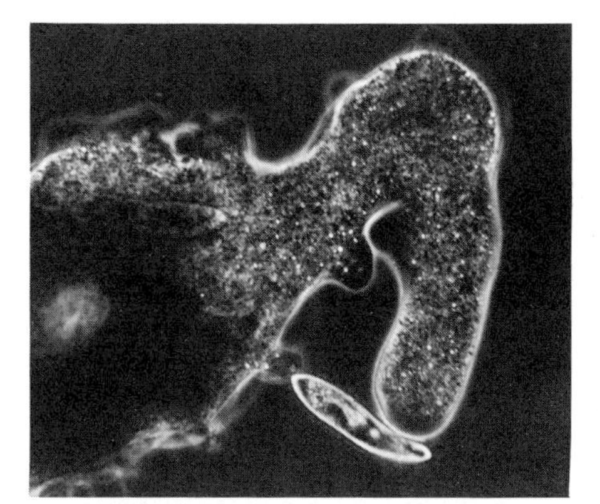

FIGURE 6-3.

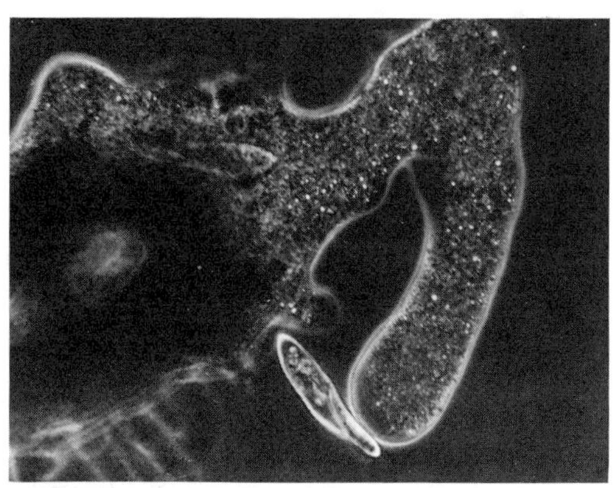

FIGURE 6-4.

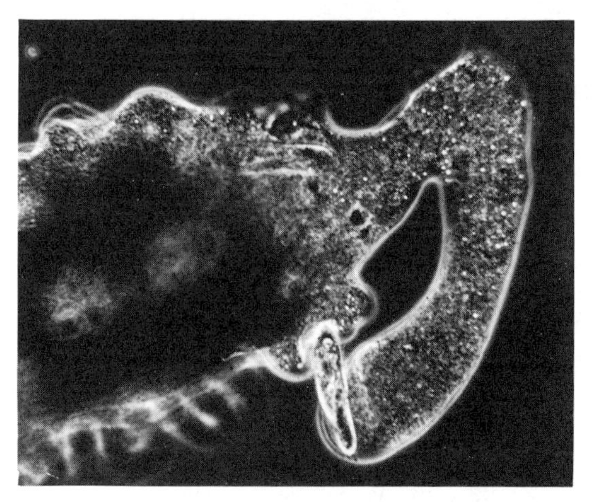

FIGURE 6-5.

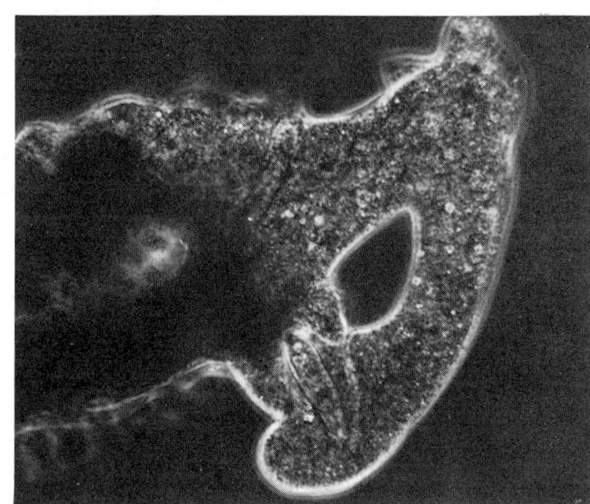

FIGURE 6-6.

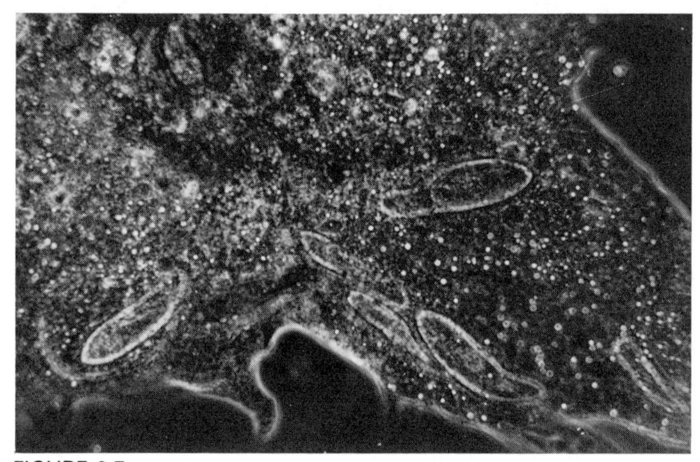

FIGURE 6-7.

FIGURES 6-1 to 6-7. See text. Figures 6-1 to 6-6, 100×; Figure 6-7, 125×.

The sluggish amoeba then becomes quick and alert. If several *Paramecia* are probing at the same time, the amoeba's pseudopods are sent out in all directions and are even capable of catching the victims simultaneously.

In the example shown in the photomicrographs, the amoeba seems to grab the prey and to direct it towards its body. In Figures 6-2 and 6-3, new pseudopods start a surrounding maneuver. Up to the phase of Figure 6-4, the *Paramecium,* which behaves in a strangely passive way, could still retreat. If the prey does escape while there is still time, the streaming inside the pseudopods ceases abruptly and is soon reversed. In Figure 6-5, however, the *Paramecium's* fate is sealed when two additional pseudopods reinforce the original one. In Figure 6-6, they fuse, forming a vacuole from which there is no escape. Yet for a long time the victim can be seen swimming in the vacuole, vainly searching for an exit.

Having ingested its fill, the movement of the amoeba's pseudopods stops. The animal devotes itself to the digestion of the meal, which may sometimes involve numerous victims. Figure 6-7 shows part of an amoeba's body in which several *Paramecia* can be seen in various stages of digestion. According to some observers, the prey, after a few hours, may be cut in half by the narrowing of the vacuole.

These photomicrographs were taken within a period of two minutes from the first contact in Figure 6-1 until the formation of the vacuole in Figure 6-6. All the magnifications are 100×, except for Figure 6-7, which is 125×.

The mode of feeding shown here is actually a very common one in nature. On a much smaller scale, it happens in our own bodies. The leucocytes, or white cells, in our blood are amoeboid cells. When we suffer an injury and subsequent infection, these cells migrate to the site of the lesion and engulf the invading bacteria in similar fashion.

Paramecium in the Robber's Den

Another giant of the microworld is the ciliate *Bursaria truncatella,* a one-celled animal living in fresh water, which reaches a size of 1 mm when fully grown—about as big as the head of a pin.

Bursaria feeds mainly on microorganisms called infusoria and wheel animalcules belonging to the clan of Rotifera. The animal has a cavity in front of its mouth called a peristome, big enough to let a *Paramecium* swim about in. The mouth opening itself—the cytostome—is actually situated deep inside the cell, at the bottom of the cavity.

Like the amoeba, *Bursaria truncatella* is helped in its feeding process because slipperanimalcules explore an object in the water by rubbing against it—the behavior known as thigmotaxis. During the *Paramecia*'s search for a suitable source of food, mainly bacteria, they often unsuspectingly swim into the robber's den. If this happens, *Paramecium* finds itself suddenly in a trap. Desperately, it swims back and forth searching for a way out. But *Bursaria*'s mouth cavity is not a simple open funnel; it is twisted, and the prey must find the exit in order to escape. I have often watched a *Paramecium*, in a vain effort to escape, constantly pushing against the same spot of the funnel wall until, by an accidental turn of its body—at last!—it finds the way to freedom. The spectacle is full of suspense. *Bursaria* can swallow the prey at any moment, and one may wonder why it hesitates. Apparently, *Paramecium* has actually to make contact with the mouth opening inside the cell in order to be swallowed. Then the victim is suddenly crushed, as if grabbed by an unseen fist, and sucked, in a flash, into the cell body.

The following picture series (see Figures 6-8 through 6-14) tells the story of the fates of two *Paramecia*.

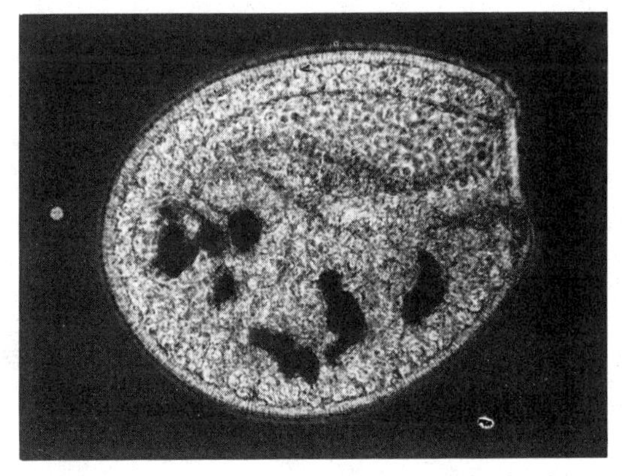

FIGURE 6-8. *Bursaria truncatella's* cell body shows numerous half-digested slipperanimalcules, 150×.

FIGURE 6-9. An unsuspecting *Paramecium* approaches . . . , 150×.

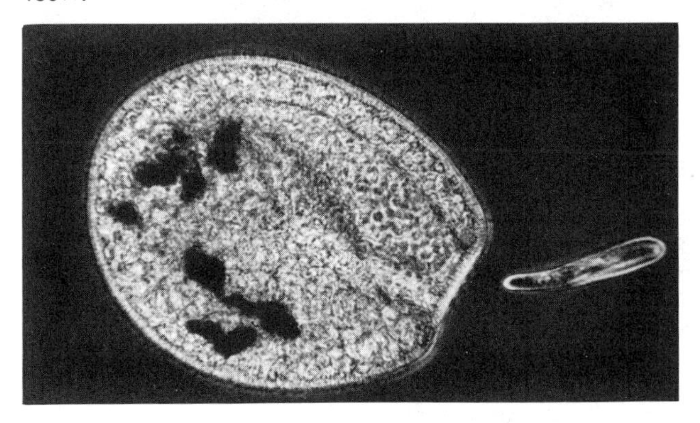

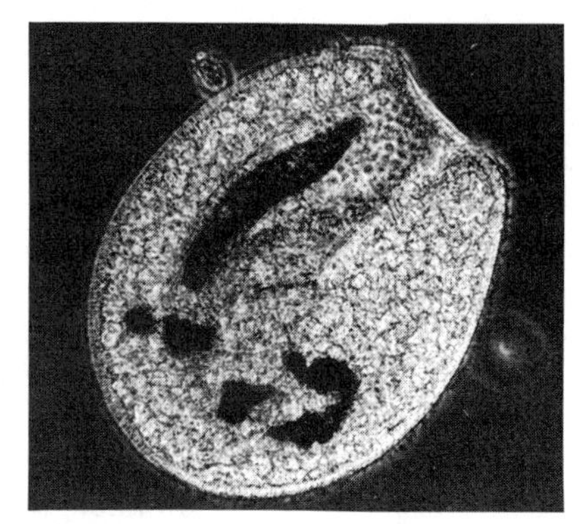

FIGURE 6-10. . . . and enters the peristome, 150×.

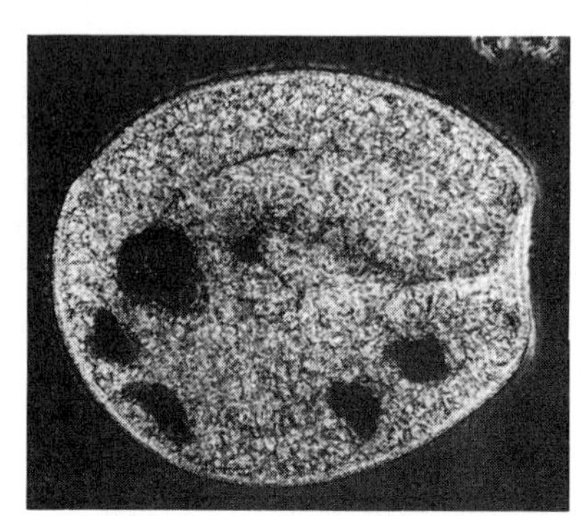

FIGURE 6-11. Suddenly, it disappears in the cytoplasm—the robber has swallowed its prey! The victim is visible in the picture as a large, shapeless blot, 150×.

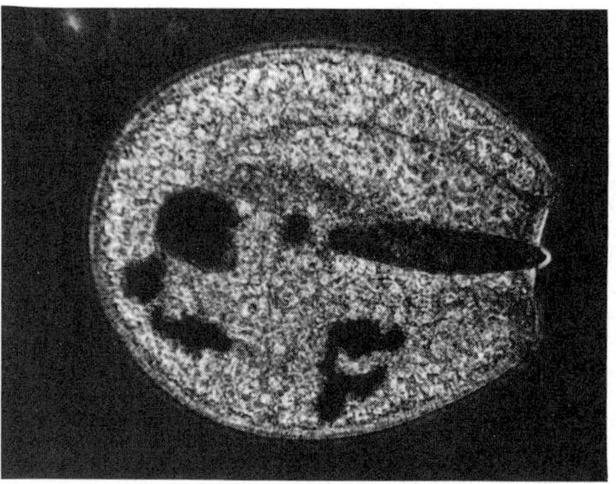

FIGURE 6-12. *Bursaria* is now ready for the next catch. A second *Paramecium* slips into the peristome, 150×.

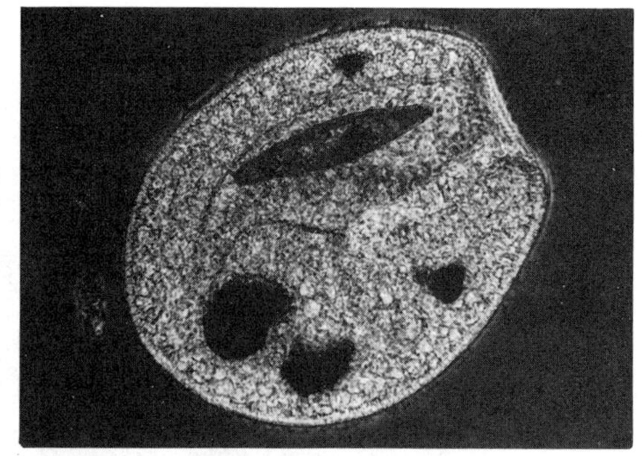

FIGURE 6-13. It swims back and forth, back and forth, trying to get out . . . , 150×.

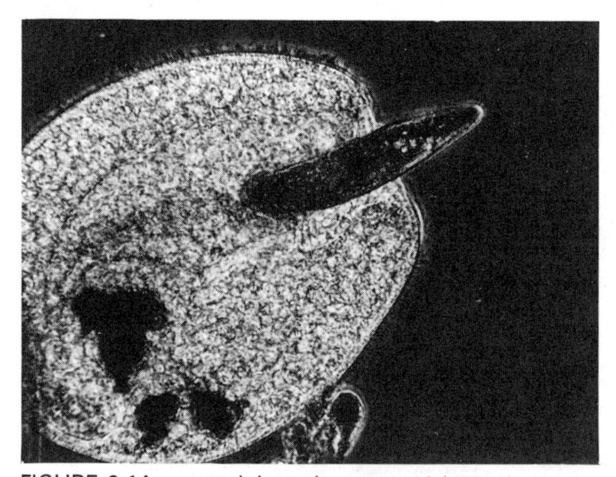

FIGURE 6-14. . . . and, hurrah, escapes! (150×)

Blepharisma, the Cannibal

A fascinating ciliate with strange habits is *Blepharisma americanum*. It is easily recognized because of its pinkish tone, an unusual coloration in the microcosm. A prominent and strong undulating membrane serves it for feeding and movement.

Blepharisma's foods are bacteria and tiny infusoria. It will swallow anything edible that can pass through its gullet. Now, it so happens that some of the denizens of pond or culture dishes develop into giant forms. This is an unusual phenomenon that puzzled early observers. One of them, Arthur C. Giese, investigating the problem in the 1930s, found that it was a question of nutrition. The giant forms are simply better fed. Giese was able to produce giants at will in his cultures by properly steering the food intake of his charges. If *Blepharismas* are fed bacteria exclusively, the development of giants is rare. But if small infusoria, such as *Tetrahymena* (a ciliate) are added to their diet, differences in size soon become apparent. Those individuals whose gullets are large enough to ingest *Tetrahymena* grow quickly. Giese proved this by starving *Tetrahymena* for 24 hours before adding them to his *Blepharisma* cultures. The starved infusoria were smaller than normal for this ciliate and, as a result, they were more easily swallowed by the *Blepharisma*. Some *Blepharismas* profited from these improved conditions to become giants. Not only did their appetites increase, but their gullets grew correspondingly, so that they were now even capable of devouring their own kind! The rich became even richer. . . .

The accompanying illustrations show this fratricidal drama. In the first figure (see Figure 6-15), the collision of a giant with a *Blepharisma* of normal size is imminent. Seconds later, the smaller animal finds itself already in dangerous contact with the giant's gullet (see Figure 6-16). Now the monster acts as quickly as lightning. Before the eye can follow, it has half-swallowed the struggling victim (see Figure 6-17). The prey is shoved into the cell body by the undulating membrane—a procedure that can be

FIGURE 6-15. The collision of a giant with a normal-sized *Blepharisma* is imminent, 145×.

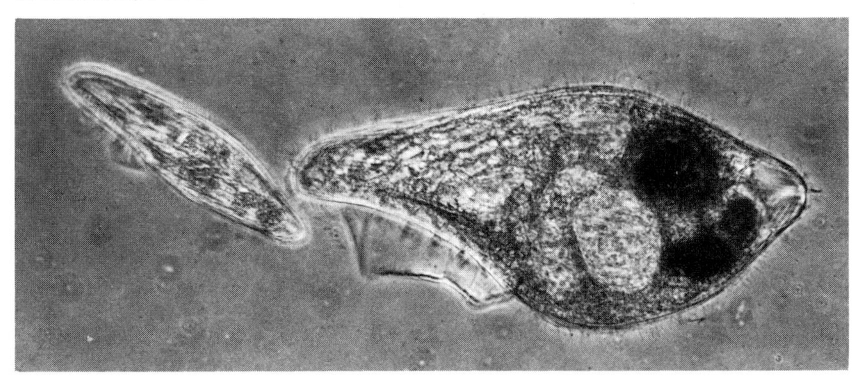

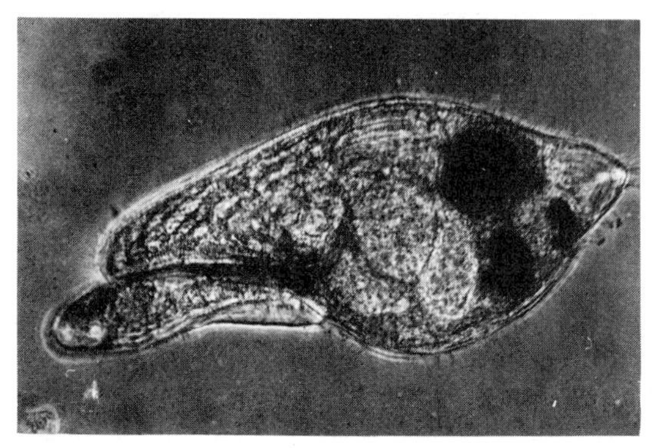

FIGURE 6-16. Seconds later the prey is grabbed, 120×.

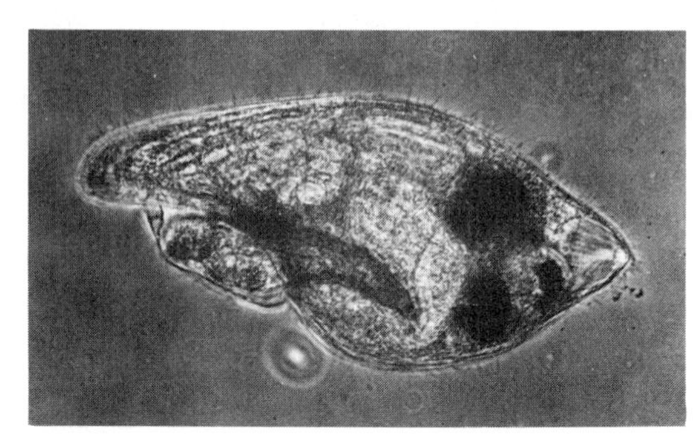

FIGURE 6-17. Now it is half swallowed . . . , 120×.

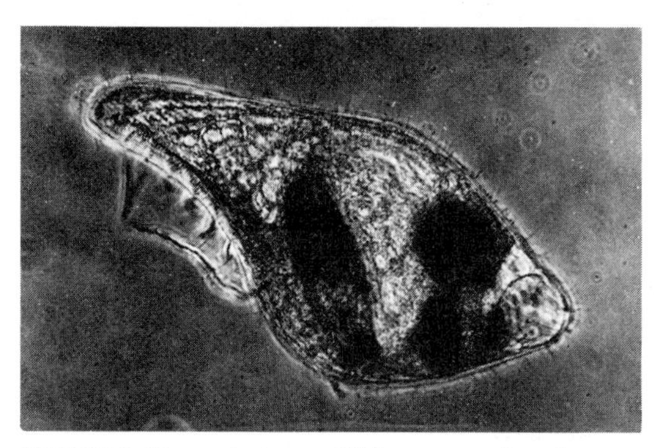

FIGURE 6-18. . . . but can still be seen swimming about in the giant's body, 120×.

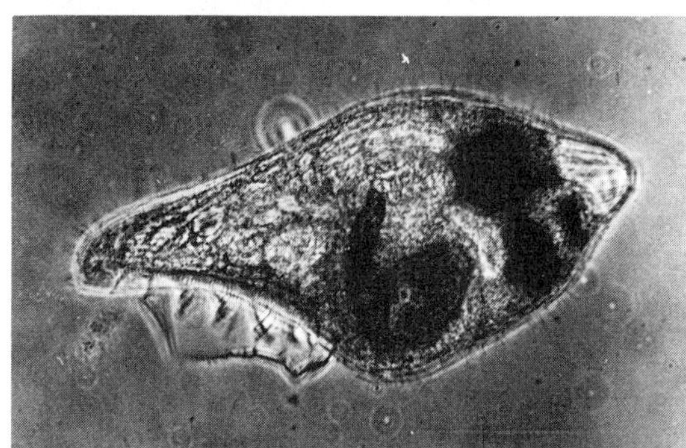

FIGURE 6-19. The narrowing of food vacuole brings about a cruel end, 120×.

very fast, depending on the size of the bite. Small animals may disappear in an instant.

In the fourth figure (see Figure 6-18), the little animal is still seen swimming about in the giant's body. Only the narrowing of the food vacuole brings about a cruel end (see Figure 6-19).

About a Unique Animalcule

As we have seen, predatory, so-called primitive one-celled animals have developed highly sophisticated methods in order to get the food they want. Snakelike, they swallow their meals whole with one bite. Yet apparently, the prey must always be smaller than the attacker. In this respect, one animal is unmatched, being able to ingest its quarry even if it is bigger than its own size. It can do it within three to four minutes. And its favored food is again *Paramecium,* this underdog of the microworld, victimized even by humans because it is such a wonderful object for research.

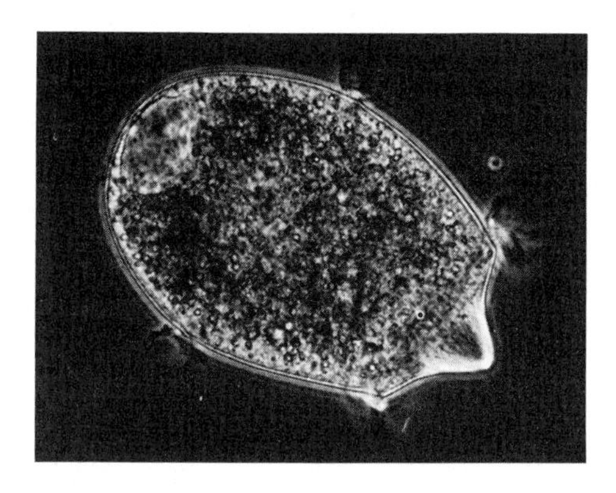

FIGURE 6-20. *Didinium nasutum,* 215×.

This unique creature is a protozoan, *Didinium nasutum. Didinium* has an average size of 200×80 microns, frequently smaller. *Paramecium*'s dimensions are 300×180 microns, frequently larger. Notwithstanding, *Didinium* can swallow the underdog with one big gulp. How does it do it?

Didinium nasutum looks like a little barrel (see Figure 6-20). It has two girdles of cilia, one running around the middle, another one around the front of the cell body. The cilia serve the animal for locomotion only. The anterior end of the "barrel" holds a conical snout or "nose," which gave this animal its species name, *nasutum.* This is the seizing organ needed to capture the prey. A contractile vacuole is located at the posterior end of the organism.

Didinium is in constant motion, making abrupt and arbitrary changes of direction. The purpose of this movement is to find something substantial to eat. It must be a protozoan, not just any one, but one with a skin soft enough to be penetrated by its snout, and there are not too many that qualify. *Paramecium* is the ideal target, but some of the smaller infusoria would also be acceptable.

If *Didinium* bumps into an object in the water, it stops and "investigates" the obstacle. Lacking any organs of detection except what we would call a sense of touch to distinguish edibles from inedibles, the probing is done by trial and error. While jabbing, *Didinium* at the same time rotates as if trying to drill into the object. More often than not, it cannot penetrate the obstruction; it then turns away, taking up the search with undiminished energy, sometimes missing a potential meal by a hair-breadth. Speed is very important. The more "microns per second" the animal can make, the greater the area searched. In a culture dish where they are numerous, they bump into one another all the time; their own skin is pretty tough, though; they soon detect the error and try their luck elsewhere. Otherwise they would be even cannibals!

What happens when *Didinium* hits a *Paramecium?* The picture series of Figure 6-21 shows this dramatic event as it occurs among living animals.

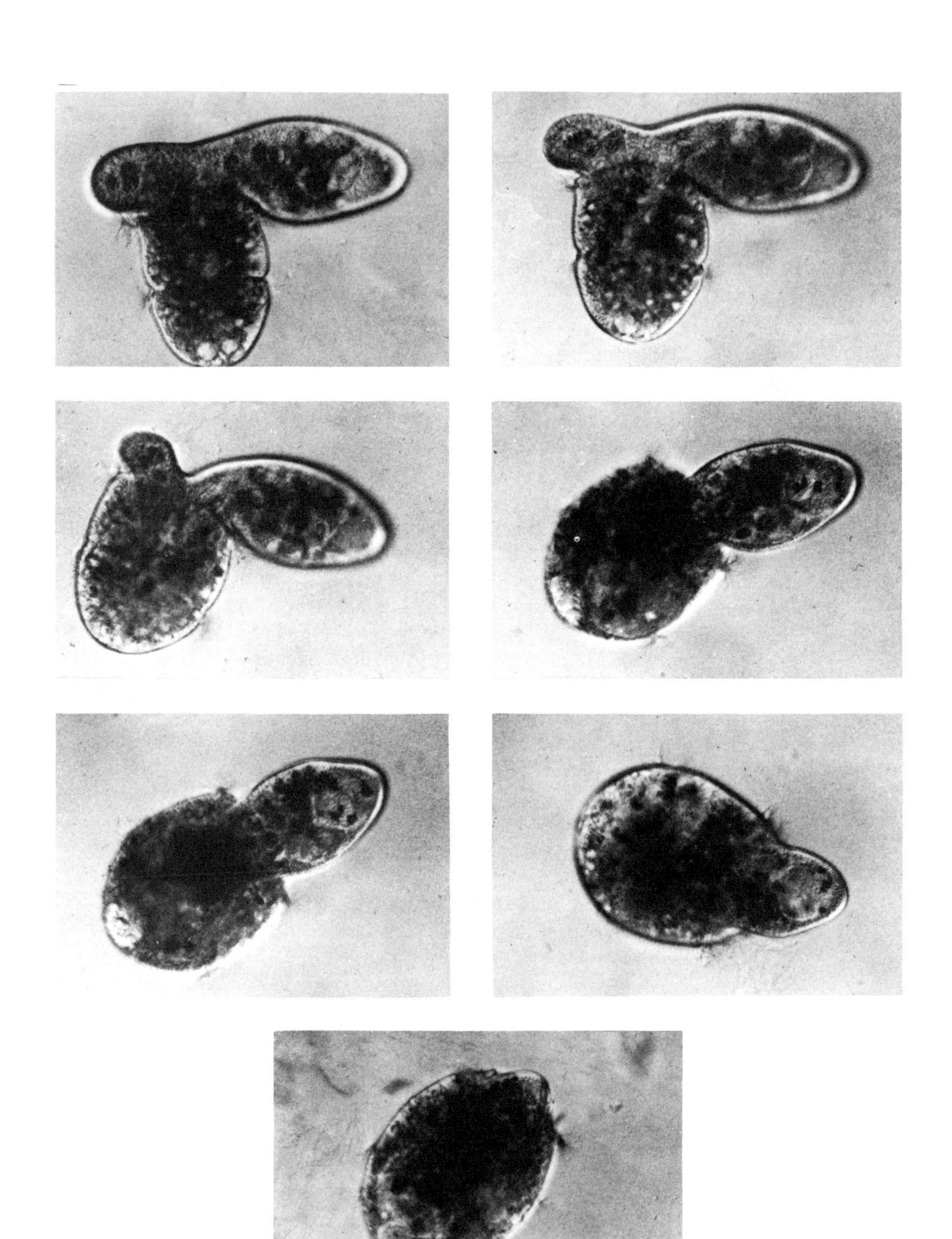

FIGURE 6-21. *Didinium nasutum* ingesting *Paramecium caudatum,* 180×.

First, the victim is pushed several times in order to verify that a suitable object is at hand. But the jabbing is also an attempt to get hold of the game. *Didinium*'s seizing organ has numerous rodlike structures, so-called trichites, which have two functions: to fortify the "nose" and to hold the prey. As soon as *Didinium* succeeds in getting a firm grip, its gullet opens wide and the seizing organ begins to move toward the posterior end of the cell body, engulfing the hapless victim. Where the force originates that pulls *Paramecium* into the cell body is not known. Is suction present? Some protozoa are equipped with myonemes, fibrils that can contract and expand, comparable to the functioning of muscle fibers. No such structures have been reported so far in *Didinium*. S. O. Mast, an early biographer of this ciliate, speaks of "inherent spreading forces" that expand *Didinium*'s cell wall, thus sucking in the prey.

An ingested *Paramecium* completely fills the predator, pushing the whole cytoplasm against the cell wall. The animal itself becomes a huge food vacuole. After the meal is digested, the seizing organ moves back to the front and is ready for another catch.

An open question is whether or not *Paramecium* is paralyzed when caught. Actually, it is very passive and offers no resistance. It is known that *Paramecium* has a mechanism that has been described as defensive. This mechanism is formed by its trichocysts, hundreds of minute rodlets that are embedded in its pellicle. Trichocysts are ejected upon certain irritations. Such a situation usually arises when *Paramecium* is attacked. The exploding trichocysts are believed to form a protective shield against the attacker. This defense, as a matter of fact, is seldom, if at all effective. *Didinium* devours its prey anyhow, unfazed. Surprisingly, however, no trichocysts were ejected by the *Paramecia caudatum* used in this experiment. A test with another species, *Paramecium multimicronucleatum* yielded the same result. Why the trichocysts misfired remains unexplained. The trichocysts reacted normally when the *Paramecia* were exposed to diluted tannic acid, a compound that never fails to explode the trichocysts, as Figure 6-22 shows.

FIGURE 6-22. *Paramecium caudatum* with ejected trichocysts, 390×.

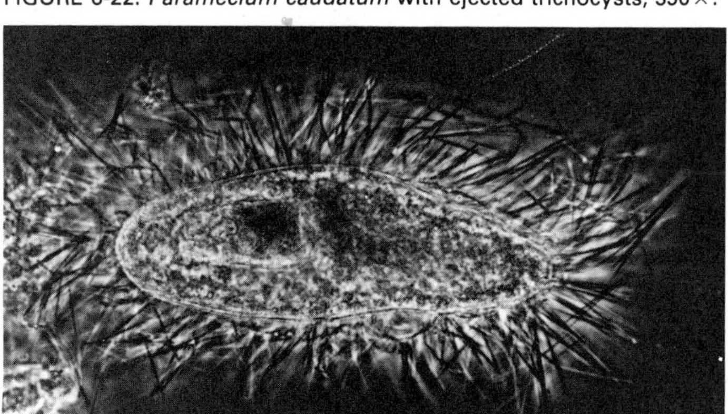

Didinium's hunting skills assure it a big chuck of food if it can find it. Under the most favorable conditions—in a culture dish in the laboratory where they are fed daily—a single *Didinium* can consume one *Paramecium* every three hours.

What happens if not enough food is present in *Didinium*'s environment? Must it die of starvation? We shall come back to this question in Chapter 7.

Lacrymaria, the Swananimalcule

Lacrymaria olor is the sonorous name of a peculiar animalcule I collected in a small headland pond. Those who know their way around in the microcosm call it the "swan animalcule." *Lacrymaria*'s neck can reach unbelievable proportions. To find this animal in a drop of water is indeed one of the most surprising experiences reserved for microscopists.

I first noticed a spindle-shaped body rummaging in a detritus flake (see Figure 6-23). I was puzzled, having never seen anything like it. Then, suddenly, something long and thin shot out of the body (see Figure 6-24), but was retracted immediately, with the speed of light. Soon the "shooting" was repeated, once to the right side, once to the left, then to the back (see Figure 6-25), always in an unpredictable direction and with unbelievable swiftness. *Lacrymaria* was busy hunting.

This animal belongs also to the ciliates. Its cell body, when resting, is about 100 microns long, considerably smaller than a slipperanimalcule (300 to 180 microns). Yet *Lacrymaria,* with its "neck" fully extended, can reach a length of 1200 microns, that is, 1.2 mm. This amounts to 12 times the size of the cell body, a truly remarkable performance. The extension is made possible by special structures, the so-called myonemes, which can be seen in some of the photomicrographs, especially in Figure 6-26, as a striation. The mouth opening is small, but can also be extended considerably.

Lacrymaria is a predator, and feeds on small infusoria. It belongs to the most elegant and agile swimmers. As it moves about from place to place, the neck is only partly stretched out and rigid; at this point, it resembles a laboratory flask (see Figure 6-27). Its speed is fantastic and can drive a photomicrographer to despair; the animal capriciously changes direction all the time; while one hand pursues it by moving the stage, the other attempts to keep it in focus and to release the shutter simultaneously.

I have encountered *Lacrymaria* only once, and then it was more or less an accident. I had collected water from a pool I had overlooked many times in the past. In the culture dish, the animals survived only for ten days. A second collection trip was not successful. Apparently, early spring is the best time to find this interesting little creature, usually in the entanglement

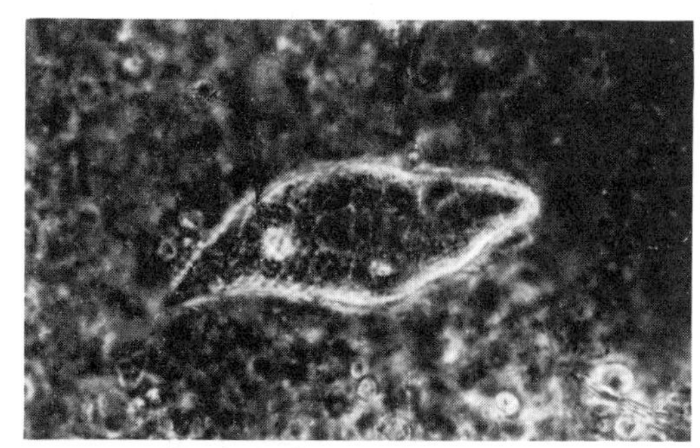

FIGURE 6-23. *Lacrymaria* contracted, 200×.

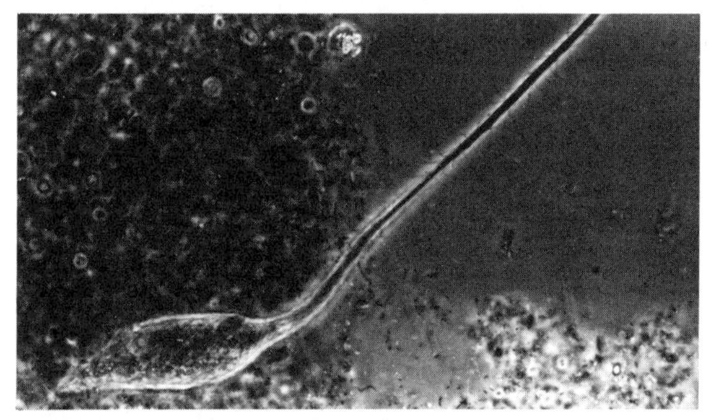

FIGURE 6-24. *Lacrymaria* with extended "neck," 150×.

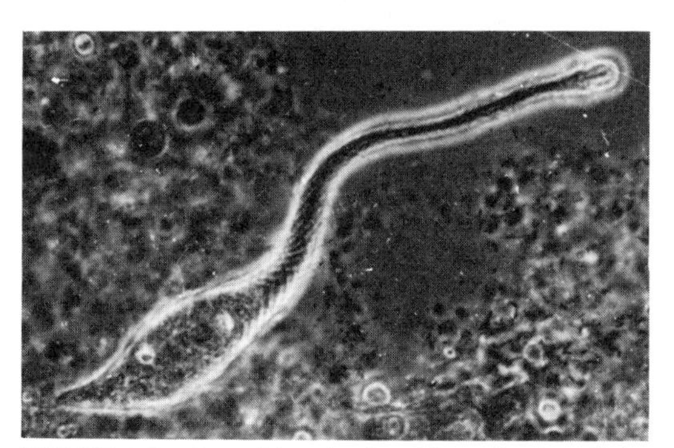

FIGURE 6-25. Searching for food in a detritus flake, 180×.

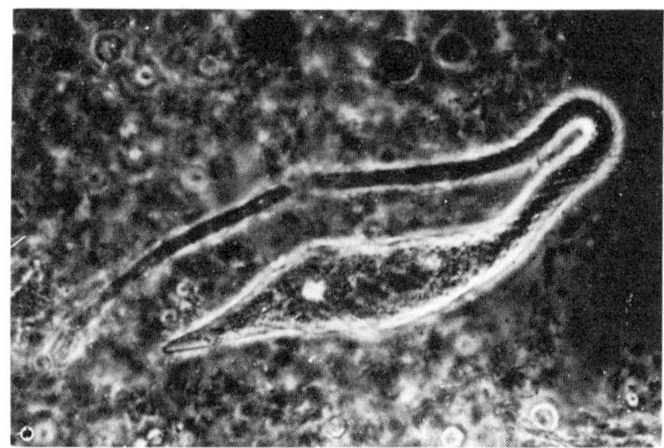

FIGURE 6-26. The partly extended cell shows the myomenes distinctly, 180×.

FIGURE 6-27. When not feeding, *Lacrymaria* navigates assuming a flasklike shape, 180×.

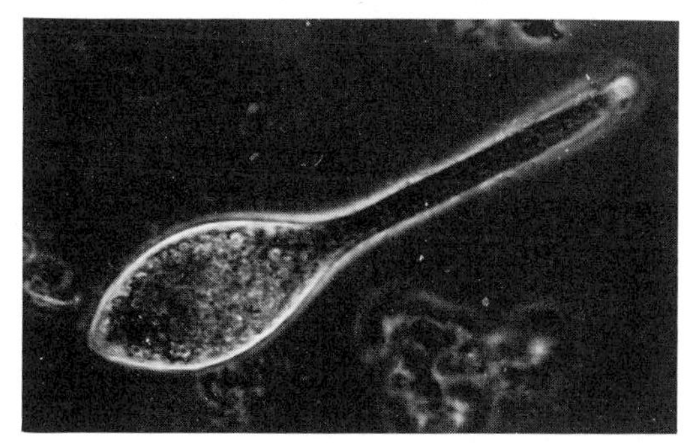

of filamentous algae. Lacrymaria likes a sheltered hiding place. It tends to settle inside the empty exoskeleton of dead waterfleas. The long neck is therefore a unique and ingenious weapon, which allows the animal to hunt without having to venture out of its safe hiding place—quite a smart way to make a living.

Hydra Catching Waterflea

One day I witnessed a drama that probably occurs constantly beneath the peaceful surface of pond landscapes: a *Hydra* catching a waterflea. The *Hydra* displayed incredible skill and cleverness. Despite its remarkable ability to contract and expand, *Hydra* is a simple animal, consisting of not much more than a hollow tube closed at one end and open at the other. The open end is the mouth and anus at the same time; whatever cannot be digested goes out the same way. Around the mouth are several tentacles—formidable weapons that serve to catch minute creatures that may happen to touch them.

Hydra has no eyes and nothing that could be called a developed brain, though it possesses nerve cells sensitive to outside influences. These nerve cells govern its activities. It is a lazy creature that moves with difficulty, and then only if it "thinks" there must be a better hunting ground elsewhere. Most of the time it hangs, head down, with outstretched tentacles and patiently waits . . . .

Maybe a *Daphnia* will pass by. *Daphnia* is a much more highly organized animal. It has well-developed internal organs—a heart, a compound eye with about 20 lenses, and a brain. It also has a digestive tract where food comes in at one end and leaves at the other. Two strong antennae enable it to move about in the water, jumping up and down incessantly. But the moment it accidentally touches one of *Hydra*'s tentacles, its fate is sealed. These tentacles are set with tiny bodies like guns, called nettling capsules, which discharge a poison that paralyzes the victim within a few seconds. For the waterflea in this series of photographs (see Figures 6-28 through 6-33), there was no escape. These photographs describe *Daphnia*'s cruel fate as it occurred.

A Plant That Captures Animals Under Water

Many people are surprised to learn that a common plant like the bladderwort should possess such an astonishing set of mechanisms with which to capture tiny animals. We hear it said that plants are so simple they need only water, air, and sunshine to get along in the world. But *Utricularia,* as

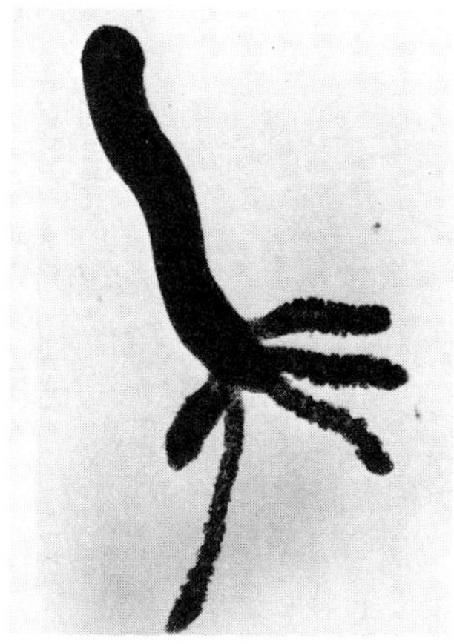

FIGURE 6-28. *Hydra* in wait for a victim, 10×.

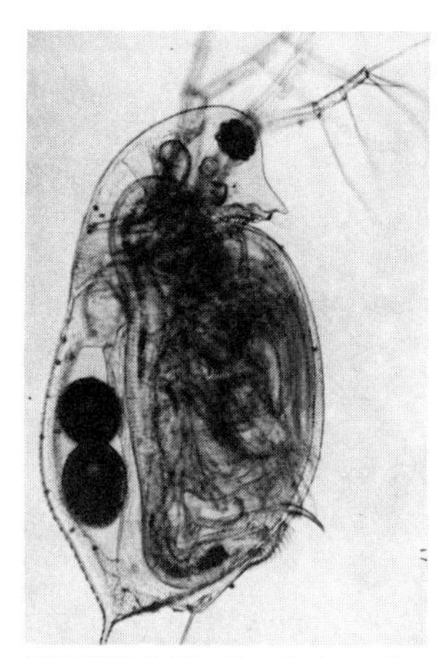

FIGURE 6-29. *Daphnia pulex, Hydra's* favorite food, 35×.

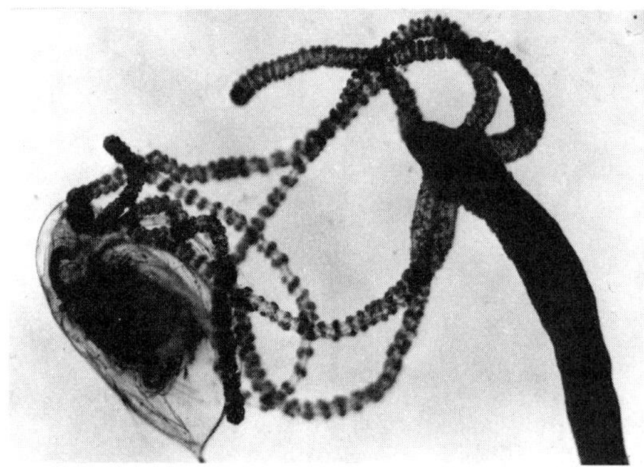

FIGURE 6-30. *Hydra's* tentacles have grasped the prey, 21×.

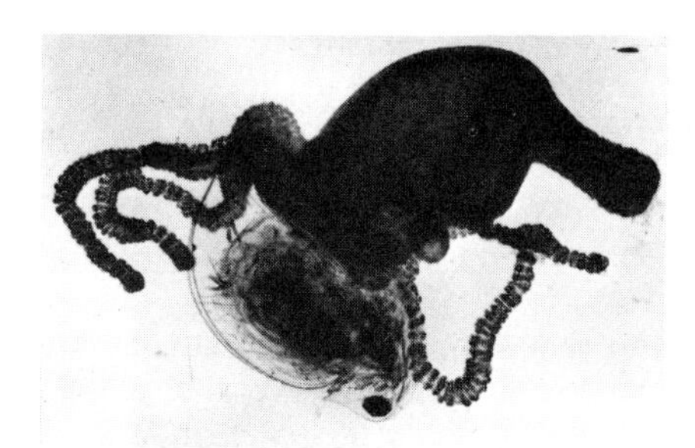

FIGURE 6-31. Its mouth opens around the paralyzed waterflea, 17×.

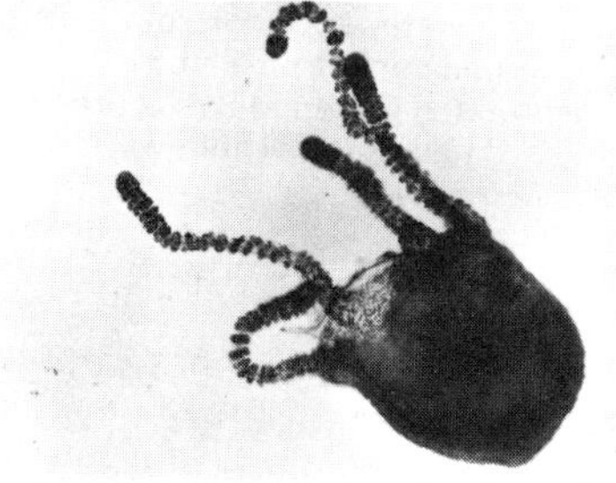

FIGURE 6-32. The waterflea has almost disappeared, 15×.

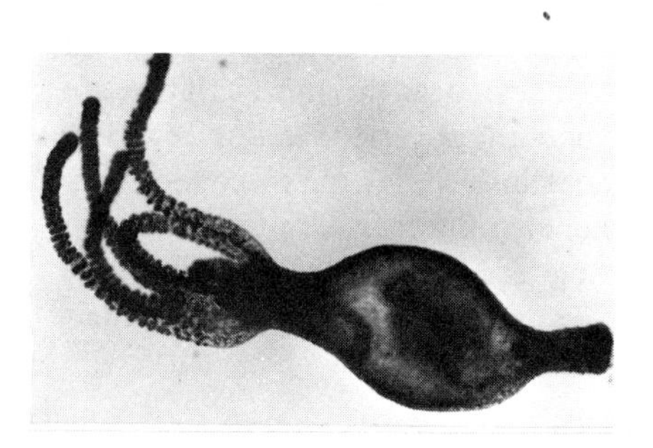

FIGURE 6-33. Now it is devoured. *Daphnia's* eye and the outline of its body can still be seen through Hydra's transparent belly, 15×.

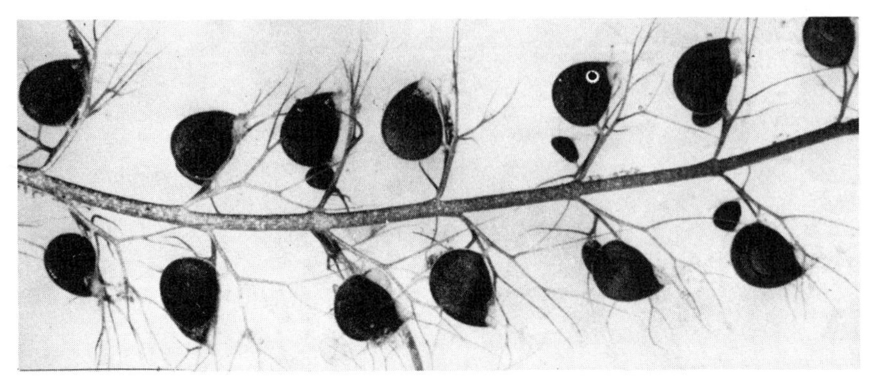

FIGURE 6-34. A group of *U. vulgaris,* magnified about seven times. The main stem of a single plant may reach several feet and support hundreds of traps.

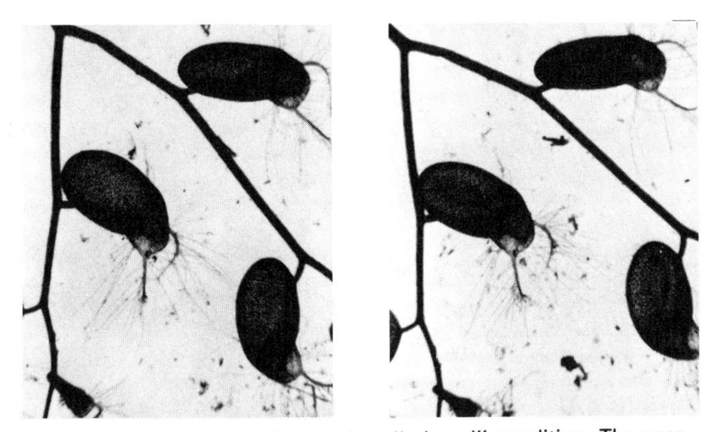

FIGURE 6-35. The traps at left are in a "relaxed" condition. The sacs are filled with water and look rounded; they are not ready to capture anything in this condition. Half an hour later (at right) the fluid has been absorbed and the sidewalls are sucked inward. The traps are now set and may go off whenever the trigger mechanism is touched, 22×.

it is called scientifically, has evolved a trap that almost gives it status as a predatory creature.

You can find the bladderwort in swampy waters or in the shallow inlets of many lakes. It is easily overlooked, though, because it floats submerged near the surface. Its pretty yellow flowers, about ½ an inch in diameter, do stick out of the water on stiff stems, but the countless traps with which the plant catches its prey are not connected to the flower. They are arranged neatly along delicate branches below the surface (see Figure 6-34). These traps look like tiny sacs with openings in front. A microscope is necessary to detect more than their general shape. A number of appendages can be seen attached to the edge of each opening (see Figure 6-35). These guide the prey toward the dangerous mouth. Woe to the animal that follows their invitation!

The mouth of the bladderlike sac is the most sensitive part of the whole mechanism. It has a "door" of fine tissue, which is only partly attached to the inner cell wall, as though on a hinge. When closed, the free edge of the door rests against a thickened pad, which forms a threshold. With a fine bristle, you can push the door in. When you withdraw the bristle, the door snaps back into position.

It was originally supposed that the bladders are filled with air to keep the plant afloat. Then, in 1875, when the German botanist, Ferdinand Cohn, first recognized them as animal-catching traps, he believed that the little animals on which the plant feeds unsuspectingly forced their way into the bladders by pushing in the door. Charles Darwin assumed the same thing when he independently tried to explain their function a year later. Today we know that the entire apparatus is a much more complicated affair.

Two things had to be discovered before the working of the trap could be fully understood. First, from four to six stiff bristles were observed attached to the lower edge of the door. These are called trichomes; when touched, they set off the trap. Second, it was shown that when the door is closed, the sac is completely watertight. Not only does the door fit snugly, but a delicate membrane, called the velum, and a mucilage excreted by glandular hairs arranged around the mouth help to seal it. This watertightness is the most essential feature of the whole mechanism.

When the plant is first pulled from the water, most of the bladders will appear rounded and bulging like well-filled purses (see Figure 6-35, left). But if put into a jar of water and left undisturbed for half an hour, the side walls of the bladders will cave in like hollow cheeks (see Figure 6-35, right). This is because the water inside has been absorbed by hundreds of microscopic hairlike structures called quadrifids (see Figure 6-37). The collapse of the walls of the bladder is possibly only because no water can seep in while the process of absorption is going on.

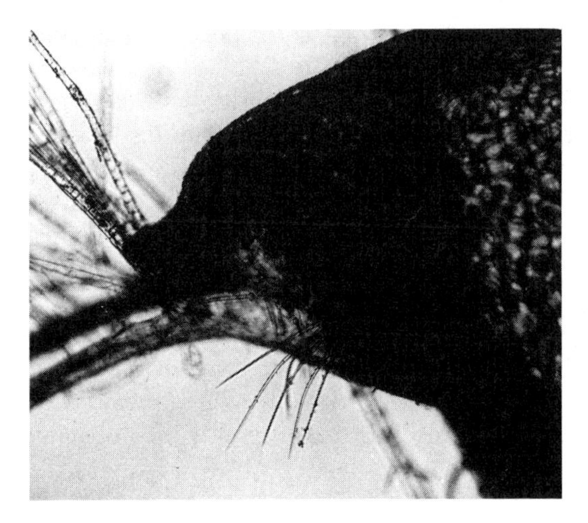

FIGURE 6-36. If any small organism touches the six bristles, or trichomes, which are attached to the door inside the snout, the trap will spring open and the incoming water will suck in the organism, 5×.

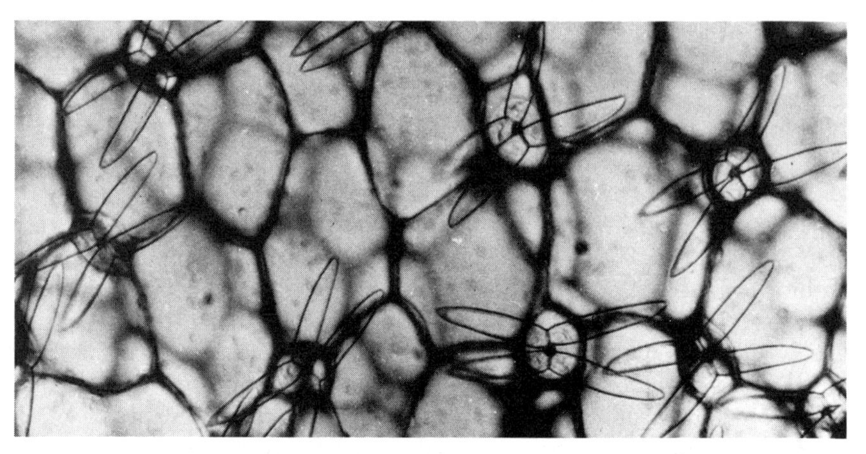

FIGURE 6-37. The cells on the inside of the sac are lined with hundreds of hairlike structures called quadrifids. Their function is to absorb the water from the trap and reset it, 300×.

In this condition, the sac is like a rubber syringe from which the air has been removed and the opening closed with one's finger. The bulb of the syringe remains collapsed, and you can feel the suction at the tip. In the same way, the *Utricularia* trap is under tension until the water can be admitted. But the door stays firmly pressed against the opening and the bladder remains in a position of delicate balance until some little animal approaches and touches it off.

Perhaps it is a waterflea (*Cyclops* or *Daphnia*) or one of the countless nematodes or insect larvae that swim and wiggle about in the water. Whatever the prey, contact with one of the trichomes in front of the door causes the edge of the door to be lifted a trifling amount. This is enough. The water pushes the door wide open and rushes into the sac as the walls spring back to their relaxed position. The unfortunate victim is sucked into the trap, and it closes immediately. There is no escape. The mechanism of an *Utricularia* trap is so sensitive that even an animal as small as a *Paramecium* may touch it off.

As soon as the trap is closed, the hairlike organs inside start to absorb the water. The side walls then fall in and things are gradually made ready again for another capture. Though the sac is less than 1/8 of an inch long, it is a big stomach with a good appetite for anything so small as a waterflea. And it means slow death for those who are swept into it. Through the microscope, one can watch the prey swimming around inside the trap as though puzzled about their strange imprisonment (see Figure 6-38).

A cruel fate awaits animals too large to be swallowed. If *Utricularia* are kept in an aquarium and provided with waterfleas, some of them may get caught by their antennae (see Figure 6-39). Unable to free themselves, they struggle desperately and finally die from exhaustion. Since opening

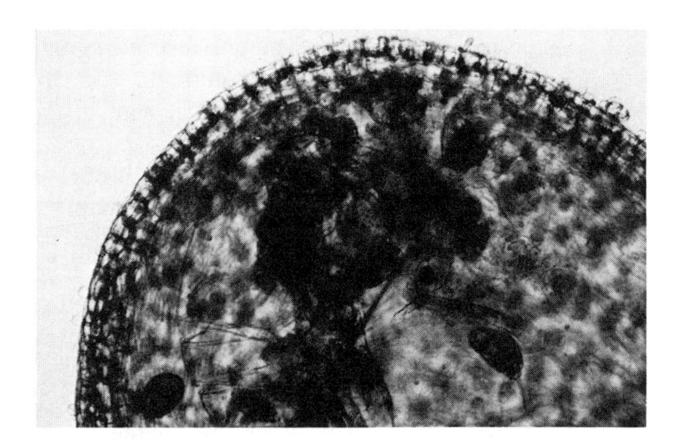

FIGURE 6-38. A view inside a trap. It contains the partly digested remains of a waterflea and other victims. At the lower right, a *Rotifer* is still alive and swimming about, 100×.

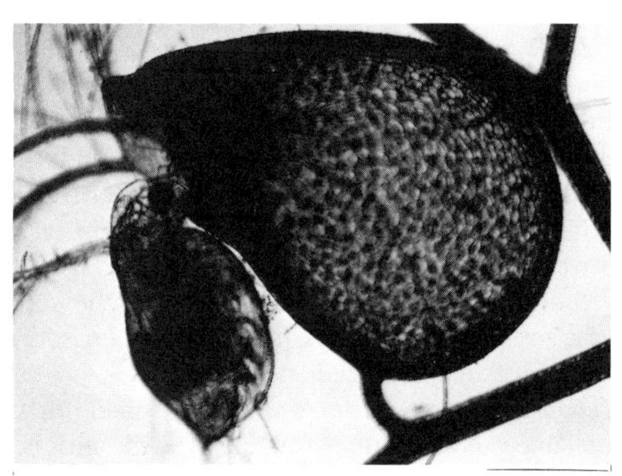

FIGURE 6-39. This waterflea was caught at its antennae. It is past help and will perish from exhaustion, 20×.

and closing of the trap takes only 1/35th of a second, there is no time for withdrawal once the release mechanism has been touched. The tentacles of the *Hydra*, which paralyze its prey, are much more merciful.

A single plant of the most common *Utricularia* in the United States *(U. vulgaris)* can have hundreds of traps. Charles Darwin believed that the animals caught in the traps decomposed, and that the decaying fluid was absorbed by the quadrifids. We now know that no decay takes place inside the traps. On the contrary, chemical analysis indicates that the quadrifids secrete digestive juices and that a process occurs that is similar in some respects to the digestive process of animals. Benzoic acid, a compound that prevents decay, has been found in the bladders; and the bacterial flora found there are sometimes types common to the digestive tracts of animals.

Some of these bacteria belong to the coli group; others are believed to take an active part in the digestive process by breaking down albumin. The chitinous or horny parts of the prey cannot be digested; they remain in the bladder and finally cause it to fall off the plant.

It is interesting to cut a trap in half and view the inside through a microscope. Cut a wedge into a light splinter of wood, just large enough to hold one of the bladders. Put a bladder into this wedge, with its mouth facing outward. Then make a cautious cut with a safety razor blade. Both halves of the bladder will stick to the blade and can be easily transferred to a drop of water on a microscope slide. This will enable you to see the hairlike quadrifids inside. You can also watch a host of microscopic organisms—protozoa, rotifers, diatoms, and desmids—some of them still alive. In fact, a bladderwort can provide abundant organisms for the microscope. The traps are at their best in spring and early summer.

The Venus's-Flytrap

Once, as I was driving south through North Carolina, I stopped at an exhibition of wild plants a few miles north of Wilmington. "Thousands of Venus's-Flytraps on View!" the billboard blazoned. What a treasure, I thought—for this is a plant that grows nowhere else in the world but on a narrow coastal strip of the Carolinas. Charles Darwin had tried to wrest the secret of this "most marvelous of all plants" from a few specimens he kept alive, with great difficulty, in a greenhouse; he was forced to gather information on it by corresponding with American friends.

The Venus's-Flytrap *(Dionaea muscipula)* is a relatively small plant whose leaves, arranged in the shape of a rosette, grow close to the ground where they can be easily reached by spiders, ants, and other insects (see Figure 6-40). It blooms in May and June and carries pretty white flowers on stiff, upright stems. The leaves end with a peculiar trap mechanism that is a masterpiece of imagination and expediency. It consists of two lobes, which—in open condition—are oriented at an angle of about 80 degrees (see Figure 6-41). The leaves have strong bristles at the rim. At the upper surface, which has a pinkish hue, six minute hairs grow, three on each lobe.

If an insect crawls over the leaf, it cannot help but touch these hairs, whereupon the leaf closes. The bristles interlock like fingers folded in prayer, to form a cage, and the insect is caught. Next, *Dionaea* starts digesting its prey. Numerous microscopic glands distributed over the surface of the leaf secrete a digestive juice to dissolve the victim, except for its hard chitinous exoskeleton. When all the useful substances have been extracted, the same glands that produced the digestive juice absorb the fluid, which is now enriched by the nutritious content of the prey. After digestion, the leaf

FIGURE 6-40. Group of leaves of *Dionea muscipula,* 2×.

FIGURE 6-41. Open leaf of *Dionea muscipula.* On the lobe turned to the camera three tiny sensitive hairs are visible, 4×.

opens again and is ready for the next catch. Each leaf is capable of catching three or four insects; then it withers, to be replaced by a younger one.

A few sophisticated details stress the unbelievable perfection of this trap. First of all, the leaf does not react if a trigger hair is touched only once. In order to close the leaf, it must be touched twice—or two hairs must be touched once each. This is a safety device to prevent the possibility that a leaf might close unnecessarily, should one of the sensitive hairs be hit by a raindrop or an object carried by the wind. The sensitivity of the leaf depends on two factors: the temperature, and the intervals between the irritations. At 40° C, a single touch of one hair can activate the trap. At 15° C, two touches are always necessary. If the touches occur within 20 seconds, the leaf responds after the second touch; if the interval is increased to one, two, or three minutes, six to nine touches are needed to close the leaf. The speed of the closing movement also depends on the temperature. On a hot summer day, a trap may literally clap shut half a second after the second irritation, if it is irritated again within 20 seconds. If the leaf is irritated only every four minutes, the closing movement not only starts much later, but also takes much longer.

A further refinement of *Dionaea*'s trap mechanism concerns the closing movement itself; the approach of the lobes after the trap is triggered occurs in stages. At the beginning, the movement goes only so far that the bristles at the leaf rim interlock, but leave a small opening (see Figure 6-42). There is a good reason for this—it gives small insects a chance to escape. It doesn't pay for *Dionaea* to start its digestive apparatus for a meager bite—an ant, for instance. After the initial closing movement, there are a few minutes until the trap closes completely (see Figure 6-43). The side walls keep their bulging shape. The final closing occurs only if the leaf caught something worth its effort; then the side walls press together, giving the leaf a flattened appearance (see Figure 6-44). This last stage requires a chemical stimulus that can be produced by any albuminous substance. The secretion of the digestive glands can also be introduced only by a chemical stimulus. If you tickle the trigger hairs with a grass blade, the trap will close completely, but will reopen within one or two days. Only if an insect "pulls the trigger" will the trap remain closed as long as is necessary in order to digest its prey—about 10 to 14 days.

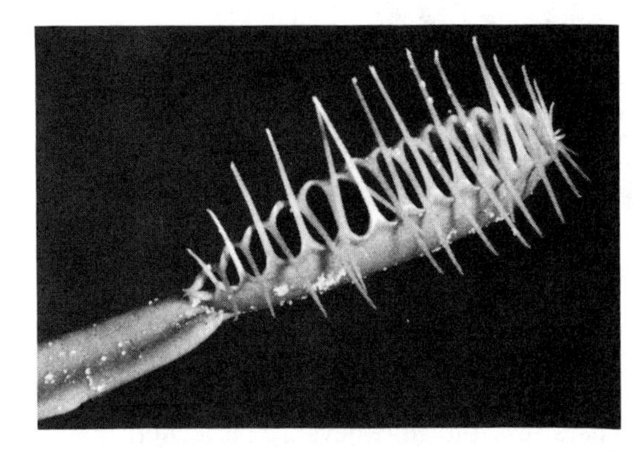

FIGURE 6-42. Leaf after the first closing movement. The bristles interlock, leaving an opening to enable small insects to escape, 4×.

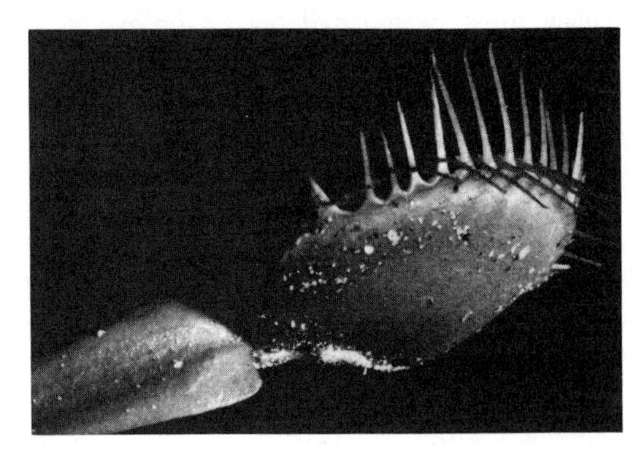

FIGURE 6-43. Leaf after the second closing movement, 4×.

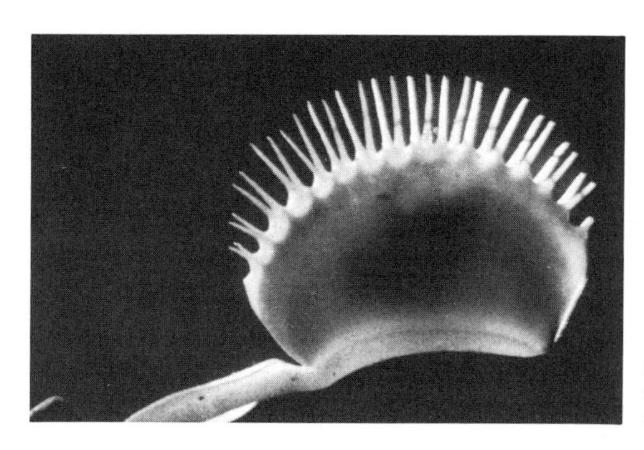

FIGURE 6-44. Closed leaf in which the silhouette of a trapped insect is visible, 4×.

That *Dionaea* sorts out insects according to size, so to speak, and rejects the small ones, was recognized by Darwin.[1] Later, in 1923, his conclusion was supplemented by the well-known American botanist, Frank Morton Jones.[2] The upper surface of the leaf, Jones found, not only has digestive glands, but also another kind of gland that secretes a juice to allure insects. These bait glands occupy a small strip at the edge of the leaf just below the bristles. Ants help themselves to this bait unpunished, because they are too small to interfere with the trigger hairs while feeding. The bait glands and the trigger hairs are arranged in such a way that only insects larger than 6 to 10 mm (depending on the size of the leaf) run the risk of touching the hairs.

Ever since the Venus's-Flytrap was first described in 1760 by Arthur Dobbs, Governor of North Carolina, it has fascinated generations of botanists. Early observers did not recognize that *Dionaea* is a carnivorous plant. Some believed that the leaf closed at night, catching only such insects as found themselves accidentally on the leaf. John Ellis thought that the trigger hairs functioned as tiny spears to spit the struggling insect, thus preventing its escape and shortening its suffering.[3] In the past 100 years, biologists have gathered voluminous material on the Venus's-Flytrap. Yet we are still far from understanding its basic mechanism.

Darwin devised an ingenious experiment to find out whether the upper and lower epidermis of the leaf undergo any changes during the closing movement. He perforated half of the leaf by cutting out a narrow strip vertically to the midrib. After the leaf had reopened, he marked its uninjured half with three small black dots, arranged vertically to the midrib at a distance of 1 mm in such a way that they could be observed through the cutout of the other half. After the leaf closed, it appeared that the distance between the three dots had decreased, while corresponding measurements on the lower epidermis indicated an increase of the distances. Darwin concluded that the upper epidermis contracted and the lower epidermis expanded during closing. Later experiments partly confirmed these observations. It was found that the upper epidermis remained unchanged, but that the lower one does indeed expand a considerable degree. When the leaf reopens, the reverse takes place, according to this interpretation.

These observations have been explained as turgidity changes. The anatomy of the leaf shows a swelling tissue of tubelike cells with thin, elastic membranes. At the instant of irritation, this tissue absorbs water from the vascular bundle system, which results in changes in the lower epidermis and leads to a bending of the lobe.

In view of the remarkable speed of this reaction, it was thought that tensions must be present in the leaf that keep it open in a state of unstable balance. "When stimulated, the balance of forces is upset and curvature immediately follows. Something happens to release the tensions. What this something is we do not yet know."[4]

Electrophysiologists have gathered interesting facts in investigating *Dionaea* with their own methods. It is well known that many animal and plant tissues produce electric currents that have been studied, especially with muscle and nerve tissues. The electrocardiograph takes advantage of these forces by measuring the action potential of the beating heart.

In 1873, the Englishman L. Burdon-Sanderson proved the existence of electric currents in the leaf of the Venus's-Flytrap.[5] According to his findings, a normal current runs from the stem to the end of the leaf. Burdon-Sanderson made his measurements with a galvanometer, whose two electrodes he connected with the leaf. It turned out that an irritation of the trigger hairs had a distinct effect on the current. If Burdon-Sanderson made a fly crawl over a leaf, the galvanometer needle responded as soon as the fly touched one of the trigger hairs. At the same time, the leaf closed. After the fly was caught, the needle vibrated each time the struggling fly moved.

Yet after decades of research work, *Dionaea muscipula* is still the puzzling "miracle of nature" as Linnaeus dubbed it. Probably more decades will pass before we understand exactly what goes on inside that leaf tissue when an insect leg unsuspectingly bumps into the trigger hairs. Despite the enormous efforts of nature in terms of genius and imagination in creating such a plant, *Dionaea* has not fared well in this world. While weeds triumph over the whole planet, *Dionaea* remains restricted to the savannahs of the Carolinas, where it lives a precarious existence under the protection of the State. Yet the authorities cannot prevent its Lebensraum (living space) from shrinking more and more.

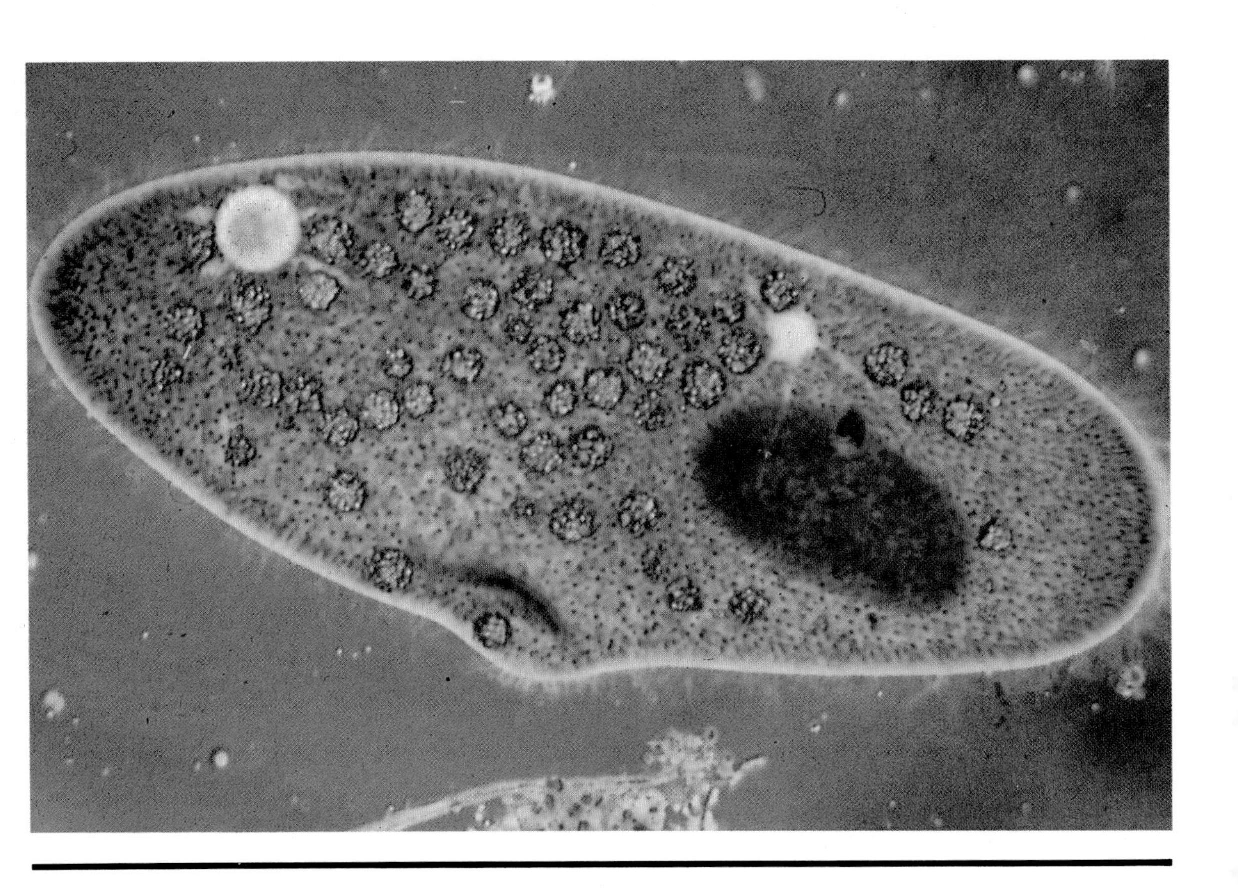

Interference Contrast

Plate 1: Paramecium caudatum *photographed with transmitted interference contrast after Jamin-Lebedeff, 440 ×.*

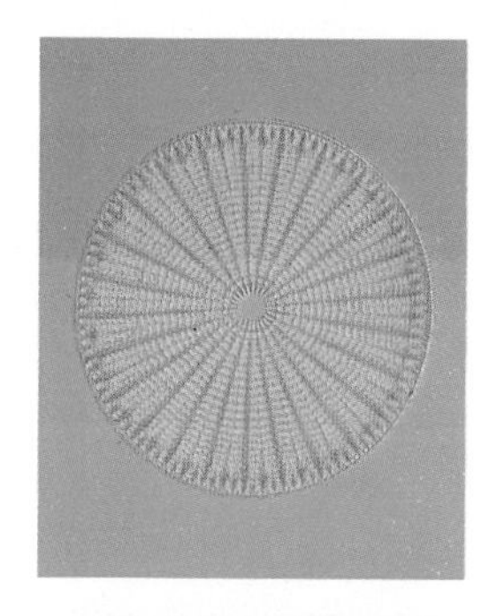

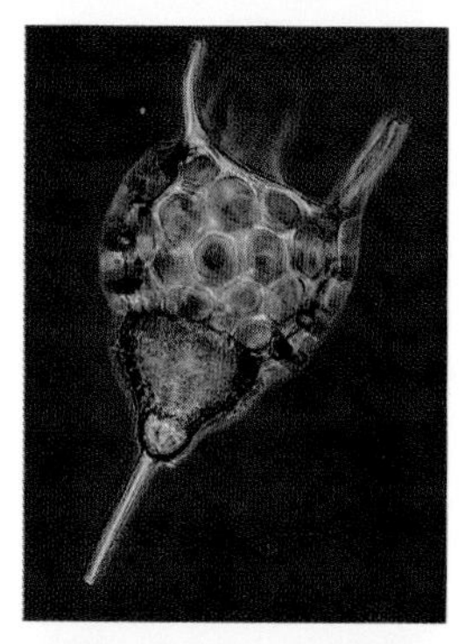

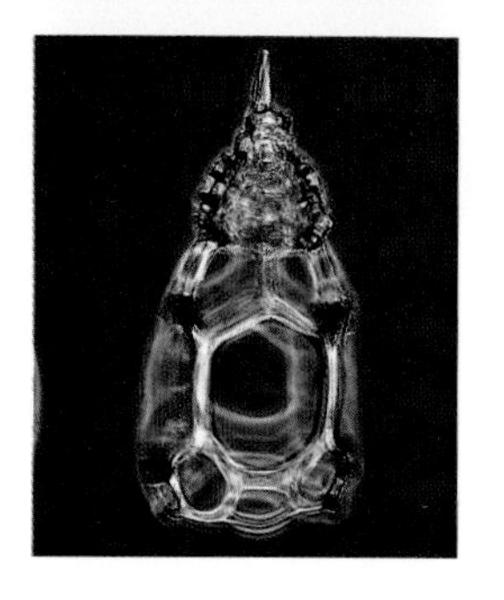

Rheinberg Illumination

Plate 2 (top): The diatom Arachnoidiscus ehrenbergi, *photographed with a two-color Rheinberg filter, the center being green, the peripheral ring yellow, 130 ×.*

The center of the filter remained black; the periphery was divided into two or three different colors. Specimens: Plate 3 (small photo in center) shows Podocyrtis triacantha, *130 ×; plate 4 (bottom small photo) shows* Cycladophora goetheana, *115 ×.*

Polarized Light

Plate 5 (large photo) shows sugar crystals under polarized light, 20 ×.

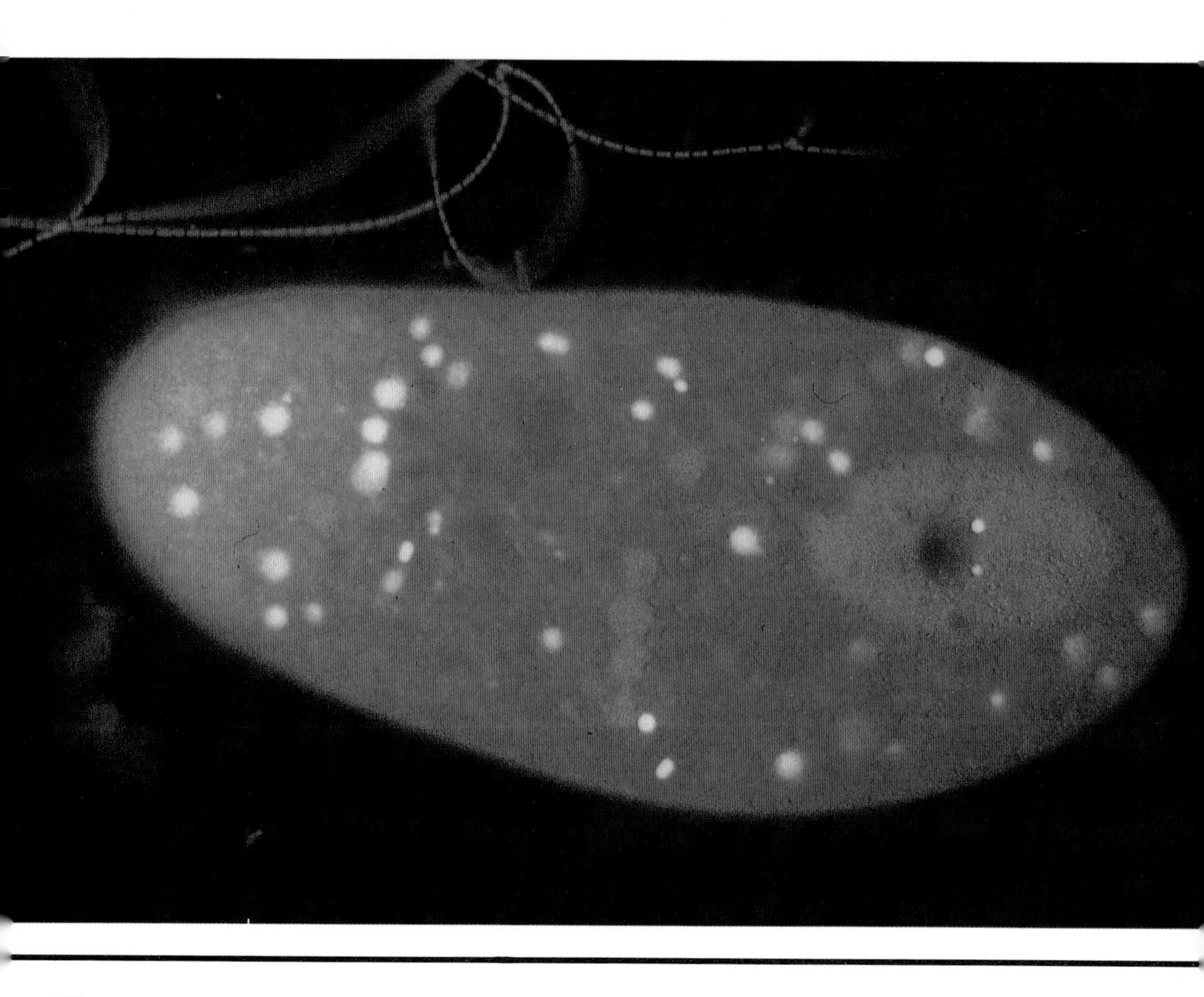

Fluorescence

Plate 6: Secondary fluorescence in Paramecium caudatum, *440 ×.*

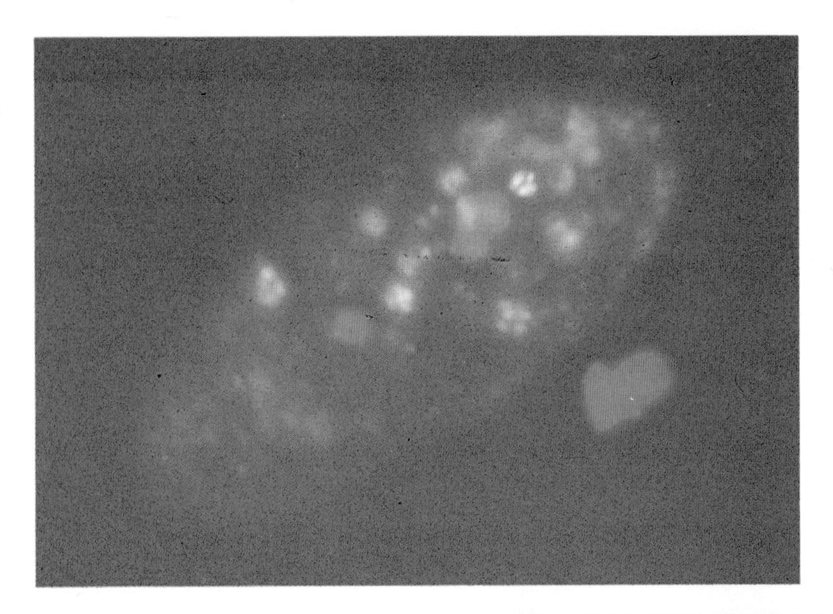

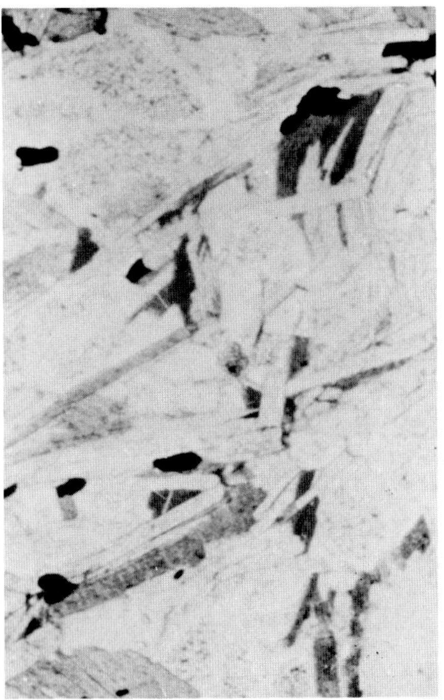

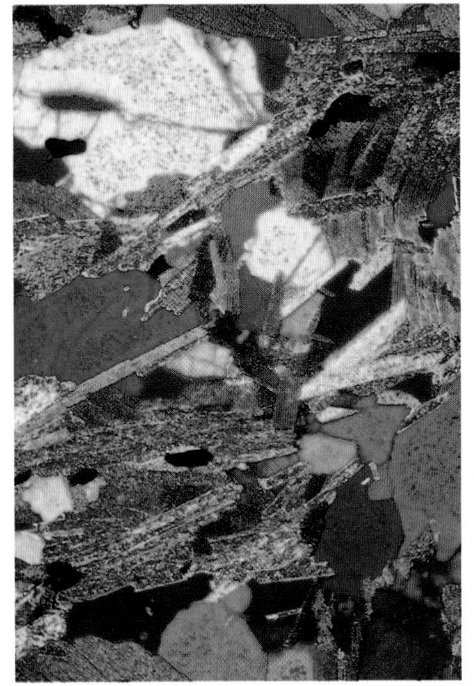

Fluorescence

Plate 7 (top): Primary fluorescence in Paramecium caudatum, *340 ×.*

Polarized Light

Plate 8A (black and white photo) shows a thin section of granite photographed with brightfield illumination, 120 ×. Plate 8B (color photo) shows the same section photographed between crossed polars, 120 ×.

7
Of Children and Children's Children

Nature, as we have seen, has provided its creatures with quite sophisticated means of satisfying their appetites. Yet merely to sustain life is not enough; the species must also be preserved. Let us look at a few examples of how this essential task is accomplished in the microworld.

An Animalcule Divides

Anyone who frequently watches the busy comings and goings in a drop of water will at one time detect a one-celled animal that is about to divide. Hastily, it passes through the field of view and is gone in a jiffy. But because we so quickly lose sight of it, we usually miss the "act of birth" itself—the dramatic spectacle when one being becomes two. Biology textbooks always show this process in schematic drawings. In nature, it is more like a violent struggle in which the two daughter cells make a mighty effort to separate. When the cell body has narrowed sufficiently, one has the impression of watching two very distinct individuals, impatient to start their own lives.

The animal shown in the photomicrographs (see Figures 7-1a through 7-1k) is a *Blepharisma americanum,* a ciliate we have already encountered as a potential cannibal (Chapter 6). It is 1/3 of a mm long, rose in color, and has a strong undulating membrane at the anterior end of the cell body that enables it to swim and feed.

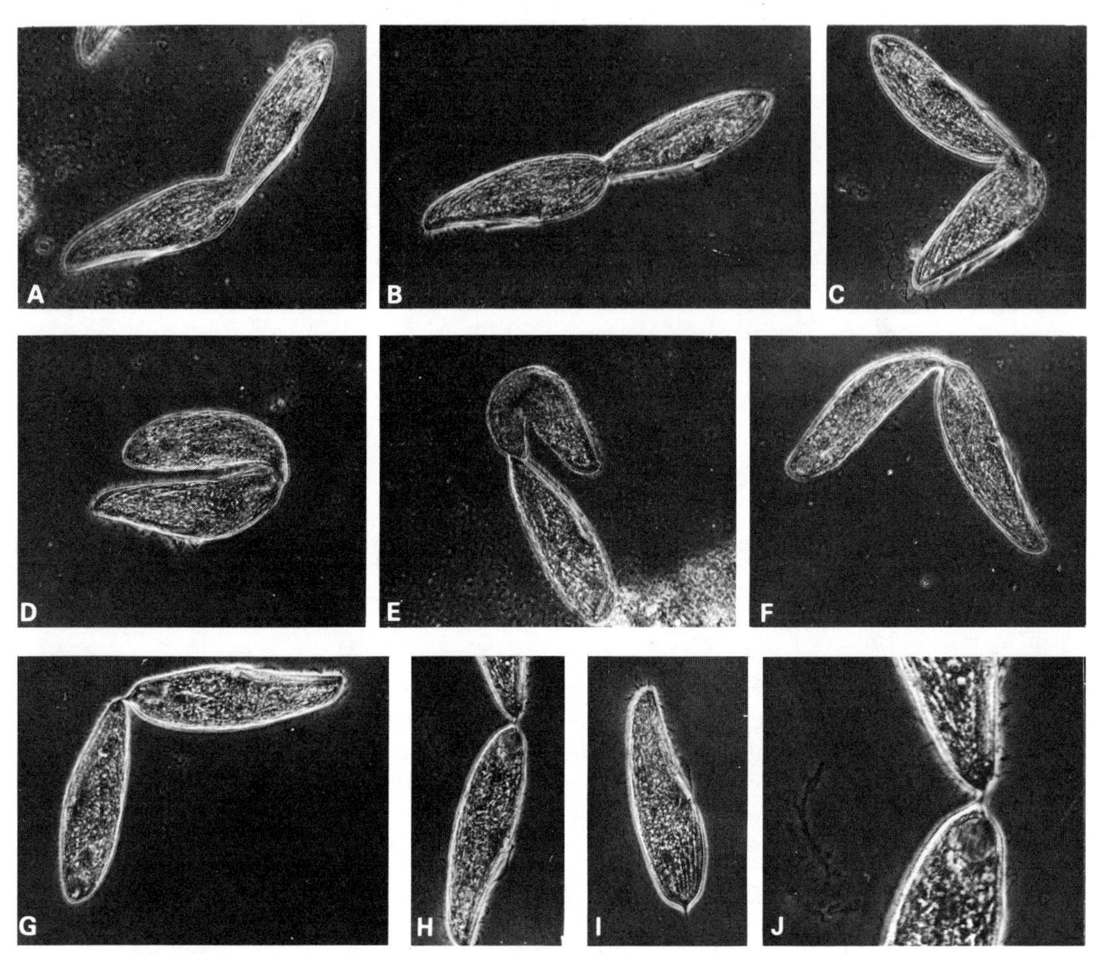

FIGURES 7-1, a to j. *Blepharisma americanum* in the process of cell division. As the division proceeds, the contraction narrows until at (h) the daughter cells are only connected with a thin protoplasmic thread, which is further magnified in (j). A fraction of a second later the cells part.
Figures 7-1a through 7-1i, 120×; Figure 7-1j, 200×.

The division of a ciliate usually takes place in two stages: first the nuclear apparatus undergoes reorganization; then the body divides. *Blepharisma* has a large, band-shaped macronucleus and a very small micronucleus, about which little is known. The nuclear reorganization starts with the macronucleus, which divides in two: one part migrates toward the front, the other toward the back of the cell. (These events cannot be seen in the living animal; to observe them, it is necessary to kill, fix, and stain it.) In the second stage, the cell itself divides as soon as the nuclear division is completed.

In Figures 7-1a to k, the various stages of the process are shown, beginning with the narrowing of the cell body. Subsequently, the contraction progresses gradually. At the same time, the organelles of the cell develop.

Each daughter cell needs (in addition to the nucleus) three basic organelles: the cytostome, or mouth opening, for feeding; the undulating membrane, for capturing food and moving about, and the contractile vacuole, for regulating the water content of the cell. The front daughter cell retains the mouth as well as the undulating membrane; the posterior cell must form these structures anew. The back cell, on the other hand, inherits the contractile vacuole, while the front one proceeds with its reconstruction. Shortly before the division is accomplished, the two cells are still connected by a thin protoplasmic thread, which breaks off after a few seconds; then the two daughter cells separate forever.

Anthony van Leeuwenhoek was the first person to observe the cell division of protozoa; failing to understand the significance, he wrote of an animal that "dragged the other one behind it." After him, other people interpreted the phenomenon in different ways. Pater Beccaria, an Italian priest, believed that the animals were mating. In a letter dated September 11, 1765, he described this to Lazzaro Spallanzani. Spallanzani, who had already seen "joined" ciliates himself, was more cautious in his interpretation, because "animalcula are a part of the creation which is still little known to the philosophers." Finally, the physician and naturalist Horace Saussure (1740–1799) found the right explanation.

Cell division is a spectacle no microscopist should miss. It requires some patience, though, to follow the process from the first indication of fission to the end without losing sight of the animal. The event shown in the illustrations took about half an hour. Details vary from species to species. In *Paramecium,* for instance, the mouth opening disappears and is newly formed by each daughter cell.

Where Quadruplets Are the Rule

An interesting variation of the "classical" method of cell division in a ciliate, just described as it occurs in *Blepharisma,* has been developed by a member of another group of ciliates, the *Colpodidae.* This is a common and widely distributed family that occurs frequently in hay infusions. But *Colpoda* are not satisfied with a simple division into two: They surround themselves with a membrane—that is, they form a division cyst—and start dividing inside this cyst into four, or sometimes even eight daughter cells.

In a flourishing *Colpoda* culture, I once found many division cysts of *Colpoda aspera* (see Figure 7-2a). They were turning constantly, like tiny globes. This rotation went on for a while in one direction, then reversed itself, then turned the other way again. The motion is explained by the fact that the cilia of the parent animal remain intact during the process of encystment and maintain their incessant flickering. But the cyst membrane stays in the same spot. This is fortunate because it helps in observing the process—you don't have to chase the cyst as you have to chase a dividing *Blepharisma*.

After a little while, two small notches appear at the periphery of the cyst (see Figure 7-2b) and soon become more pronounced (see Figure 7-2c), indicating the start of the fission. Then another notch becomes visible: the division of the first division (see Figure 7-2d). The cyst grows more and more rugged looking. In Figure 7-2e, the outline of the "embryos" becomes more distinct. Then comes the great moment: The thin membrane bursts, liberating four daughter cells (see Figure 7-2f), which immediately take off in all directions. At "birth" the daughter cells are smaller than the mature animals, but within an hour or two they reach normal size (see Figure 7-2g) which for *Colpoda aspera* is 30 microns. Each of the four descendants will encyst after a few hours and again produce four descendants, and so on, as long as living conditions in the pond, ditch, or culture dish remain favorable. Under normal conditions, *Colpoda* reproduce at intervals of about eight hours. The process shown in the photomicrographs lasted for an hour and a half altogether, from the detection of the rotating cyst until the bursting of the membrane.

If food is getting scarce—*Colpoda* feed on small bacteria—or if their environment dries out, *Colpoda* can form "resting" cysts. In addition to the thin cyst membrane, a second, more stable cyst wall forms so that the animal can survive adverse living conditions for many years, waiting for better days. It is these resting cysts that are revived if a handful of unboiled hay is dipped into a jar of hay infusion or some suitable medium and let stand for a few days. A resting cyst, as a rule, produces only one individual.

Didinium's Survival

At our first encounter with *Didinium nasutum,* when we were spectators at its dinner table (Chapter 6), a question arose about what would happen to this protozoan if the food supply were exhausted. Would it die? The answer was postponed, but can now be given: It would survive the same way *Colpoda* does, namely, by taking advantage of the ingenious solution nature provides for many protozoa if environmental conditions deteriorate. Protozoa encyst to overcome hard times and excyst when things look better again.

There are various reasons, aside from dietary considerations, why protozoa change to a dormant state temporarily. Those living in puddles are in permanent danger of drying out; or the temperatures drop or rise with disregard to their likings. Most of them must survive the winter months. Some, like the *Colpodae,* last for many years in dried condition, while *Didinium* must remain in a watery environment.

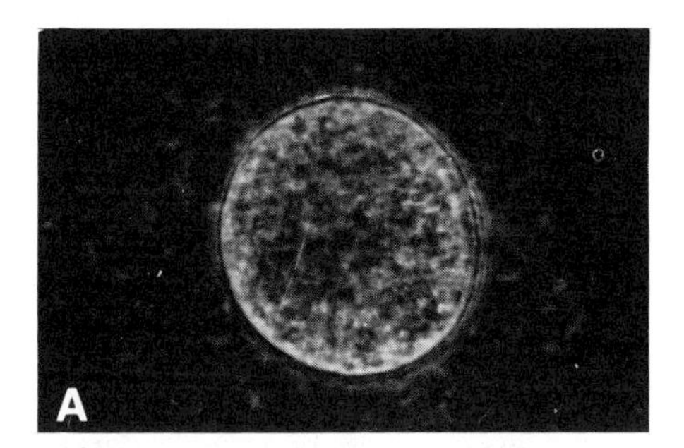

FIGURE 7-2a. Division cyst of *Colpoda aspera*, 500×.

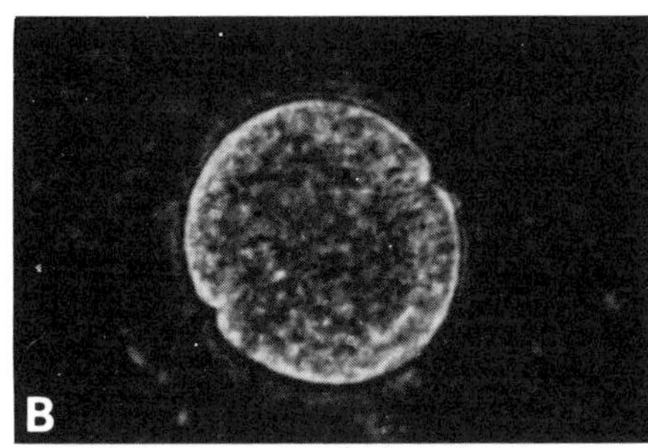

FIGURE 7-2b. The cyst shows the first cleavage . . . , 500×.

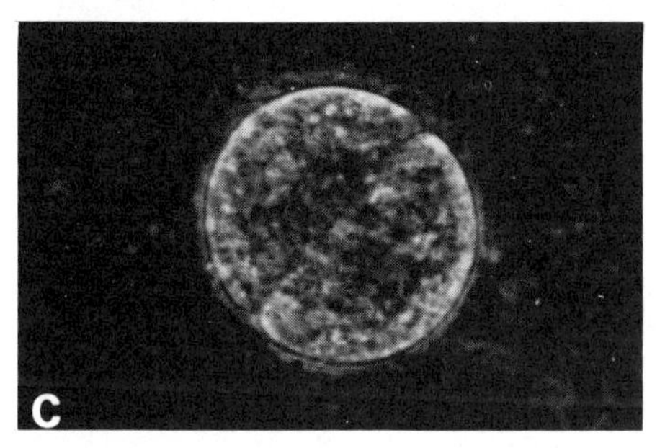

FIGURE 7-2c. . . . that becomes more pronounced, 500×.

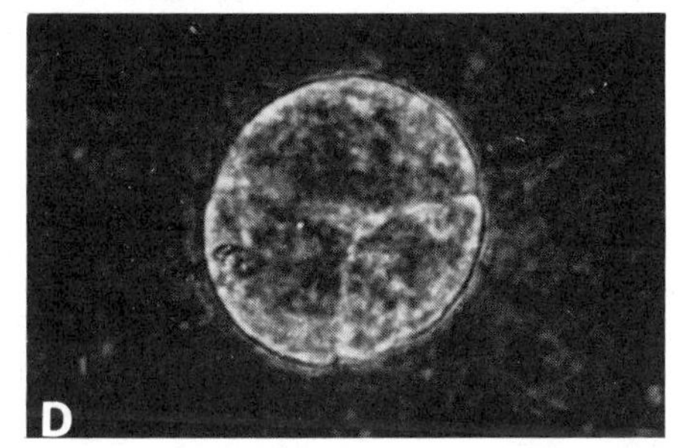

FIGURE 7-2d. A second cleavage develops, the division of the first division, 500×.

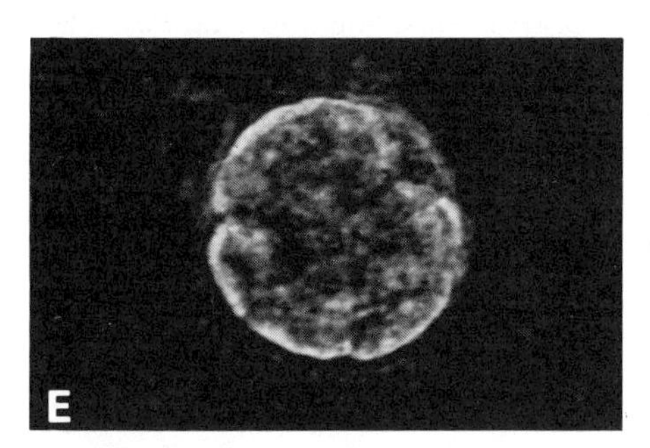

FIGURE 7-2e. The outlines of the embryos become more distinct, 500×.

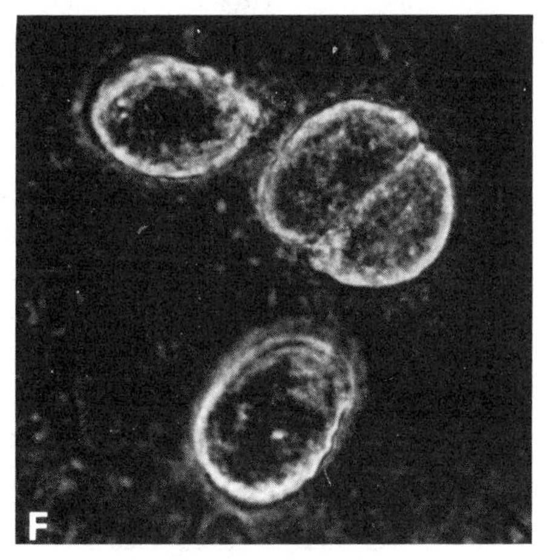

FIGURE 7-2f. The membrane has burst, liberating four daughter cells that disperse in all directions, 500×.

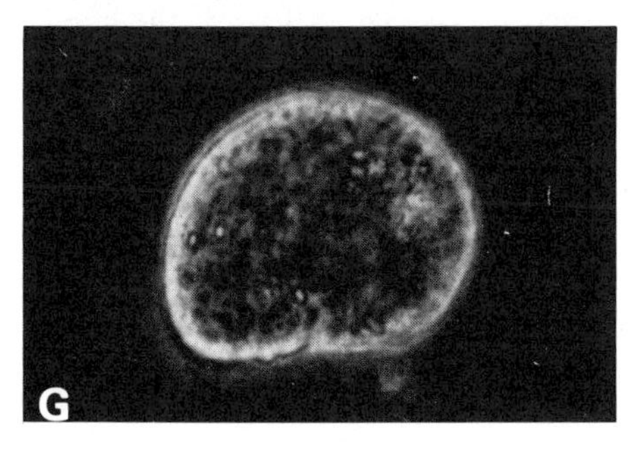

FIGURE 7-2g. Each daughter cell soon reaches the normal size of the parent cell, 500×.

For *Didinium nasutum,* the alarm sounds when the food supply runs out. *Didinium* then rounds up and forms a protective shell in which all organelles are dedifferentiated; that is, they lose their structural specifications and function. Nuclei, contractile vacuole, snout, and cilia enter a seemingly disorganized granular cytoplasm. The cyst's shell is made of three layers: two outer ones that provide protection; the inner third one, called the endocyst, is not visible in the resting cyst. It plays an important role later on as the excystment proceeds (see Figure 7-3a).

It is the endocyst that gives the observer the signal that excystment is imminent. This cyst consists of a thin membrane that first becomes apparent as a small clear spot inside the resting cyst (see Figure 7-3b). This happens when, due to climactic conditions such as rising temperatures in spring, life in the ponds becomes more numerous. Bacteria are the first to appear. Their presence changes the chemistry of the water. For the cyst, it means that conditions have become more favorable, and that the time has come to hatch.

The endocyst, which is elastic, takes in water. It slowly grows in size to a point at which it eventually breaks the protective layers mechanically (see Figures 7-3c, d, e). After first protruding through the two outer walls, the endocyst slowly squeezes itself out until it is suddenly expelled (see Figure 7-3f). This process takes several hours, depending on the outside temperature. Right after release, the endocyst discharges excess water and shrinks markedly in size within a minute (see Figure 7-3g). The cyst content condenses. Already a distinct space between the cytoplasm and the cyst wall can be detected (see Figures 7-3h to k). At this point, the embryo, looking rather shapeless, slowly begins to rotate. The rotation is a sign that cilia are forming. And, as more and more cilia are differentiated, the rotation accelerates. From here on, the endocyst gradually widens (see Figure 7-3l), giving its imprisoned inhabitant more and more space to race excitedly back and forth. Now its problem is to break out of the endocyst. It achieves the liberation by producing an enzyme that dissolves the endocyst, a remarkable wonder of microchemistry. The embryo, now rotating at top speed, constantly pushes against the cyst wall (see Figure 7-3m).

Here we watch the birth of a creature that is its own obstetrician. The observer of this spectacle, which lasts from 15 to 20 minutes or more, must wonder about the tremendous energy stored in a single cell. Will the embryo accomplish the breakthrough, or perish in its prison? Finally, the endocyst becomes uneven (see Figure 7-3n), dissolves, and disappears (see Figures 7-3p and q). *Didinium* then suddenly escapes, but just can be caught by the camera. In some of the photomicrographs, the animal itself is not sharp because the focus is on the outline of the endocyst, a large globe in which the embryo moves up and down and sideways, frustrating attempts to get both the ciliate and the cyst wall sharp. The whole series could be done only in an open dish because the cysts do not perform under a coverglass.

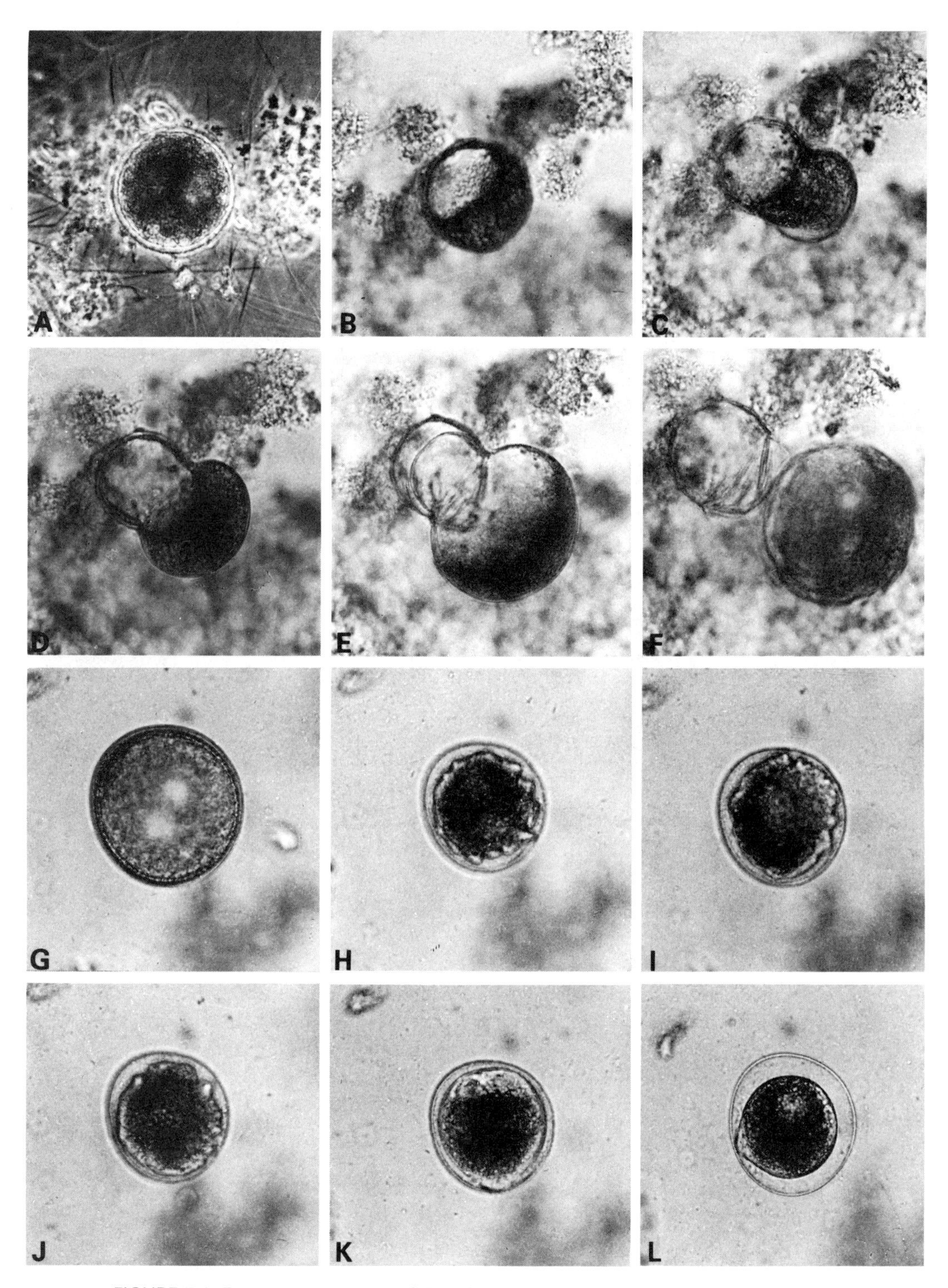

FIGURE 7-3. Excystment process of *Didinium nasutum*, 200×.

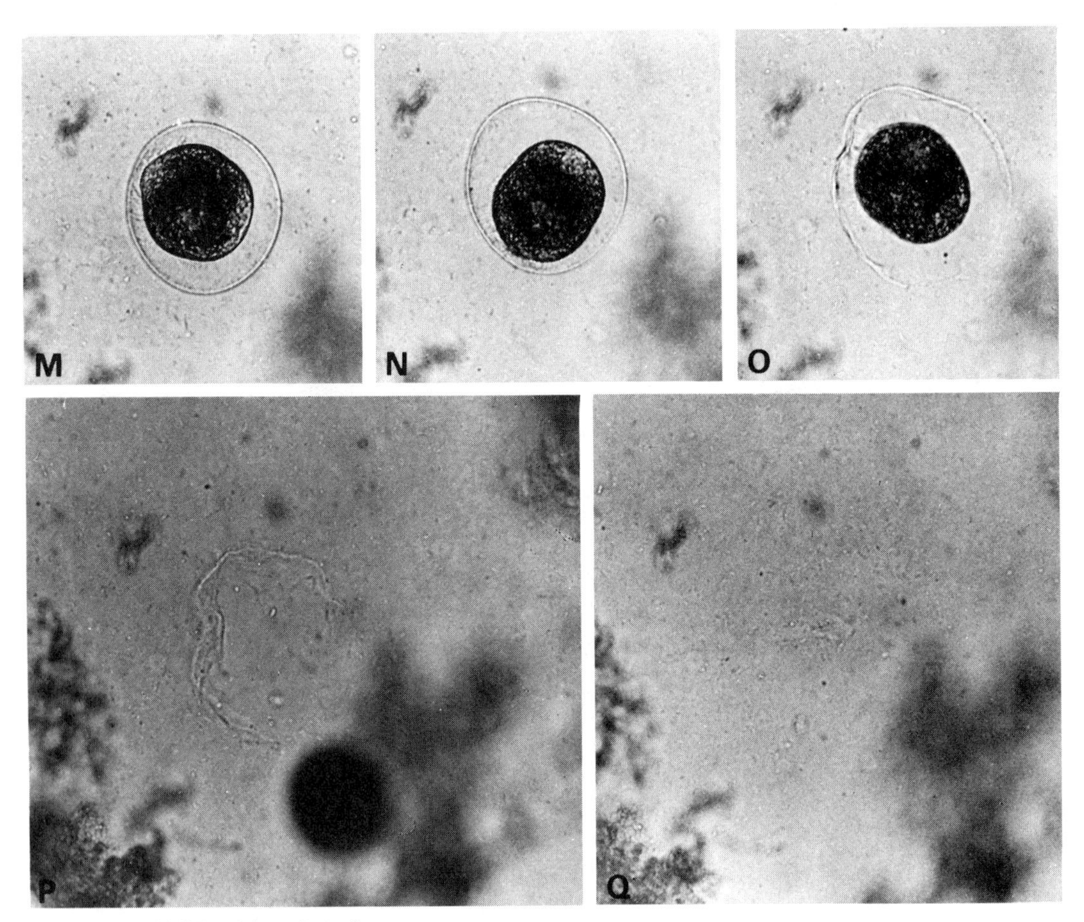

FIGURE 7-3 (continued).

The newborn *Didinium* is still immature. The seizing organ is, as yet, undeveloped. It will take another hour before the animal can resume its hunt and replace the energy spent during the excystment drama.

The en- and excystment processes are often described as if the encysted animal were dormant, comparable to the hibernating state of certain mammals. In the hibernating bear, the animal's metabolism is simply slowed down. Upon awakening in spring, it is the same animal that takes up its activities. Not so in the en- and excysting protozoa. In the encysting *Didinium,* all organelles are dedifferentiated; the cytoplasm assumes a granular consistency, the cilia disintegrate, the contractile vacuole disappears, and the nucleus loses its characteristic shape. The genetic material, however, is preserved, and when the cyst hatches, the organelles are redifferentiated. What emerges is therefore not the same individual. It has only inherited the genetic material from the preceding organism. The whole process is designed to preserve the species. It is a form of delayed reproduction for unicellular animals that live under hazardous conditions.

The viability of resting cysts is amazing. It has been established that *Didinium* cysts can survive for ten years, provided they do not dry out. Cysts from other ciliates, however, remain able to reproduce even if dried. The excystment process of different species varies too. Some protozoa hatch directly from the cyst. Others undergo rather complicated successions of events similar to those shown for *Didinium*.

The described process lasts about five hours, during which time the observer is nailed to the microscope. Counting from the appearance of the vacuole (see Figure 7-3b) until liberation, redifferentiation only starts at the stage shown in Figure 7-3h and lasts only for half an hour. What we can see in this short period of time is, so to speak, a tirelessly rotating "mixer" in which the molecules of seemingly chaotic material arrange themselves in a prescribed, methodical way, ending up with a new living being, a true replica of its ancestry—indeed an admirable spectacle.

Conjugating Paramecia

Conjugation is a process in which two individuals of the same species unite temporarily at their mouth openings with the purpose of exchanging part of their nuclear material. After the exchange, the conjugants separate. This is a process of sexual reproduction that contrasts with the asexual or vegetative reproduction by cell division, as shown in *Blepharisma* (see Figure 7-1). After conjugating, the parting conjugants then divide with renewed genetic material. It is a way of rejuvenating the race. The procedure is very complex and cannot be observed in the living animal. By fixing and staining many conjugating specimens in different stages of conjugation, it has been possible to reconstruct the events. For *Paramecium bursaria* (there is much variation in different species, even of the same genus) this has been done in the most detailed way by T. T. Chen.[1] and others. Based on their findings, the sequence of this phenomenon, described briefly and with much simplification, is as follows.

Paramecium bursaria has two nuclei, the same number as *Paramecium caudatum* (see cover picture), a macronucleus, and a micronucleus. First the micronucleus in each conjugant undergoes repeated divisions, during which one half of the micronucleus degenerates and the other half continues to divide. This happens three times. The results are two micronuclei: one is called the stationary, the other one the wandering micronucleus. The latter micronucleus got its name because it starts migrating through the mouth openings into the opposite conjugant, thus accomplishing the exchange of nuclear material. After the exchange, the wandering nucleus fuses with the stationary one in each of the two individuals. The product of this fusion is called the synkaryon. The conjugants now separate, each going its own way. However, the task has not been completed yet. The synkaryon continues to divide in several stages until four nuclei are produced. These nuclei develop into two macronuclei and two micronuclei, whereupon each post-conjugant divides. Each daughter cell gets one macro- and one micronucleus. The end result is four new *Paramecia,* the same number that could have been achieved if each conjugant had divided vegetatively in the first place, with the important difference, though, that the offspring made in this complicated manner are more youthful and vigorous.

Early observers of the microworld tried to understand what was going on during this process. But as long as conjugations in their cultures were rare events, there was never enough material on hand to reconstruct what was happening. In the 1890s, the French protozoologist, E. F. Maupas, discovered that he could induce conjugation if he brought together *Paramecia* of the same species, but from different locations; he interpreted this to mean that only partners from different "ancestors" would mate.

Maupas' observation, which he published in 1889, was later confirmed; only the terminology has changed. Today, the protozoologist does not speak of "ancestors," but of "clones" and "mating types." A clone is the total of all descendants from a single individual—from a single *Paramecium,* for instance—isolated from one natural source (such as a pond or ditch). Members of the same clone do not conjugate. Maupas' first search for conjugants in his culture dishes was futile because all his slipperanimalcules belonged to one clone. It was the American biologist, T. M. Sonneborn, who, in the 1930s, clarified the conditions that lead to mating in *Paramecia.* He obtained *Paramecium aurelia* from different parts of the United States and started clones from each of these collections. He then combined the members of different clones and examined their mating reaction. The result was that different clones do not necessarily conjugate. Sonneborn found that he could divide his clones into three varieties. Members of one variety did not conjugate with members of another variety. Within a variety, however, he found two clones, each of which did conjugate. He called them "mating types."

Paramecium bursaria (shown in Figure 7-4) is a common species of slipperanimalcules, easy to culture if the dishes are kept in a window without being exposed to direct sunlight. This is the green *Paramecium* that lives in a symbiotic relationship with the one-celled green alga, *Chlorella.* Both have established a kind of mutual housekeeping in which they exchange nutrient material. If the algae are given favorable living conditions—sufficient light for their photosynthetic activities—the *Paramecia* can be kept alive for a long time.

Figure 7-5 shows a number of *Paramecium bursaria* swimming about in a leisurely fashion looking for a suitable source of food. Each animal is

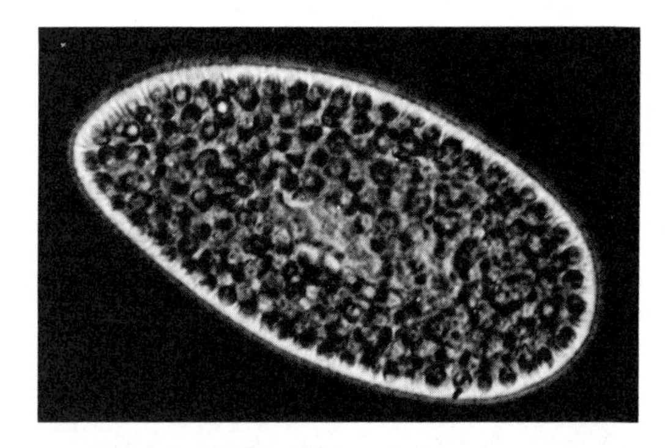

FIGURE 7-4. *Paramecium bursaria* with its symbiotic *chlorella* algae, 500×.

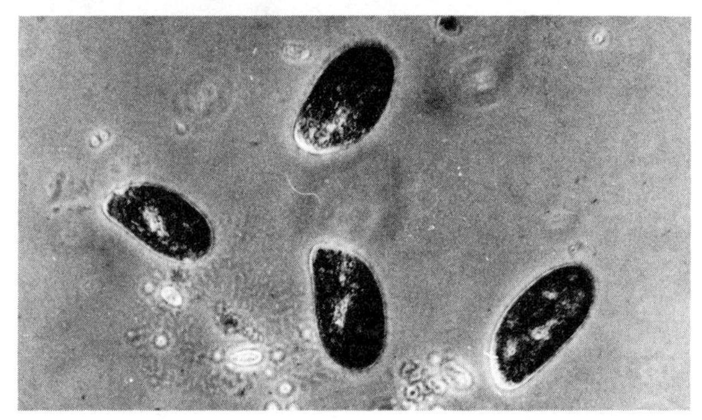

FIGURE 7-5. *Paramecium bursaria* from a normal culture. The animals remain separate, 125×.

apart from the others. They may, if their paths cross, touch one another accidentally and do indeed bump into one another frequently, only to separate again.

But if we put *Paramecia* of different mating types together on a microscope slide, a dramatic spectacle is presented: There is barely time to focus the microscope before pairs appear (see Figure 7-6). While the animals were rather lazy in their separate dishes, life now comes to the drop of water. The animals seem excited. Not that they attract one another; on the contrary, they sometimes try to continue in different directions, which makes it look as though they are performing a kind of love dance. Yet those *Paramecia* that accidentally bump into one another stick together—if they belong to different mating types—and do not separate again. They are not yet joined at the mouth openings, but at random points of contact. Figure 7-7 shows such a first encounter. It also demonstrates that the two partners are "chained together" by their cilia; there is still space between the cell bodies. Soon a third animal joins the two (see Figure 7-8) and then a fourth (see Figure 7-9). Finally, the groups grow into clumps, which sometimes hold as many as 100 individuals together. The animals remain in such clumps for up to 38 hours (see Figure 7-10). Their movement is impeded, but not

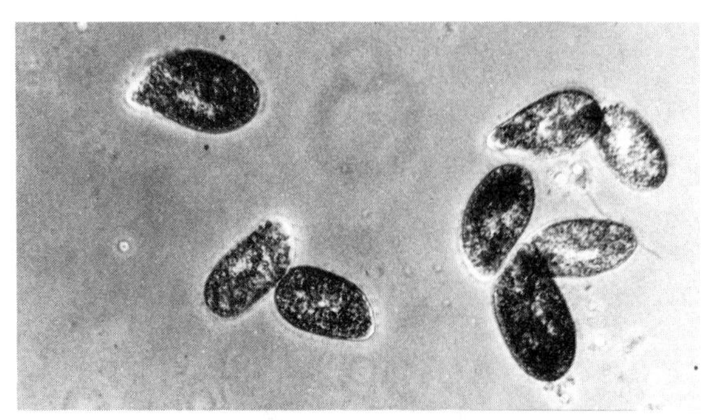

FIGURE 7-6. Animals from two different mating types are put together on a slide. They immediately join at the points of contact, 125×.

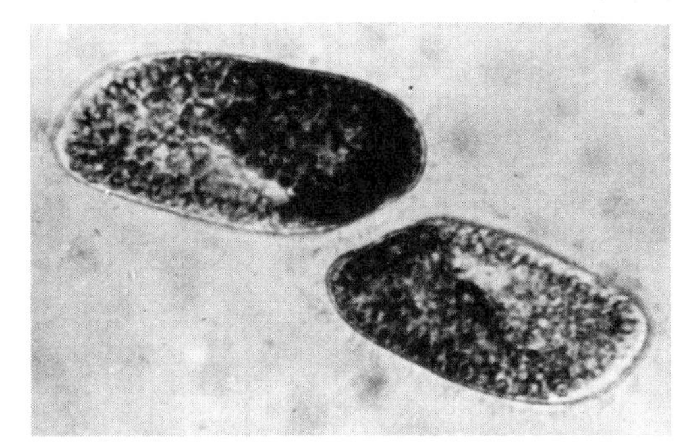

FIGURE 7-7. A pair is held together at their cilia.

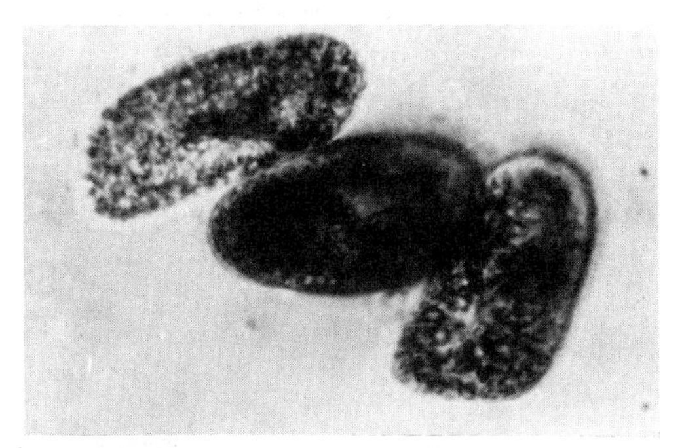

FIGURE 7-8. A third joins the pair . . . , 250×.

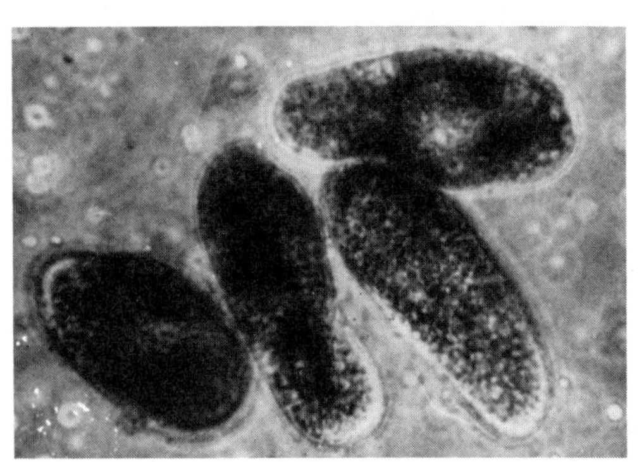

FIGURE 7-9. . . . and a fourth, 250×.

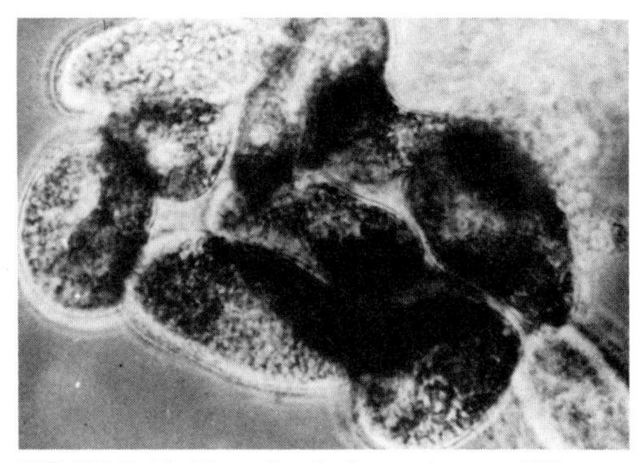

FIGURE 7-10. They finally form clumps, 250×.

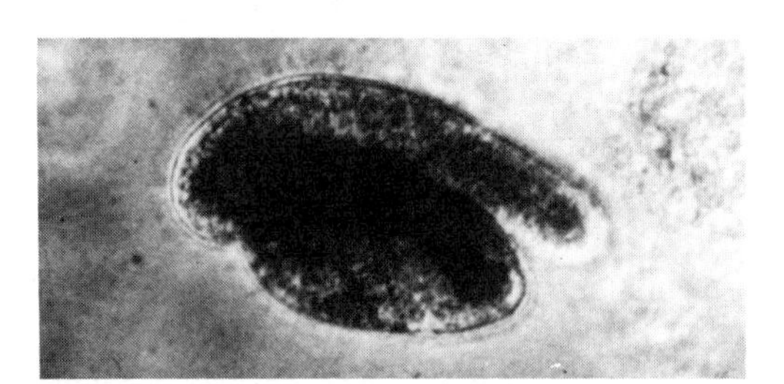

FIGURE 7-11. These clumps dissolve after several hours, liberating conjugating pairs still joined at their mouth openings, 250×.

so much that they cannot change their position in relation to one another. After about one or two days, the clumps dissolve. Numerous conjugating pairs emerge (see Figure 7-11), as do a few "lone wolves" that were left without partners.

Conjugation is a phenomenon known only in ciliates and suctoria. In the ciliates, mating types have been identified only for a few species. Among the *Paramecia,* mating types are now known of the more common species such as *Paramecium caudatum* and *Paramecium multimicronucleatum.* In *Blepharisma americanum,* on the other hand, I have frequently seen conjugation, although the cultures originated from the same natural source.

The clones, groups, varieties, or mating types of a *Paramecium* species all look alike; no anatomical differences can be detected among them. Yet they behave differently. What is it that guides these "simple" animals? What is it that brings them together? Scientists are still searching for answers to such questions.

Cyclops *in Production*

If you take a jar of water from any pond, ditch, or lake and hold it against the light, there is a good chance that you will see little dots moving back and forth, up and down. Among them there are likely to be some little animals which, when viewed through a hand lens, seem to be especially active and fast-moving. They are *Cyclopses* (see Figure 7-12), little creatures with one eye, named after the terrible one-eyed giant of the Greek saga who gave Ulysses so much trouble on his way back from the Trojan War.[2]

Cyclopses are very prolific animals—they have to be in order to survive because they form an important link in the food chain. They feed on protozoa, diatoms, and desmids, but are themselves food for all kinds of small fishes.

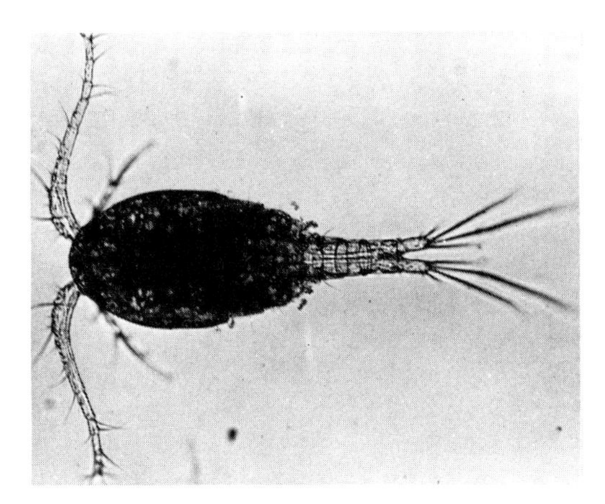

FIGURE 7-12. A young female *Cyclops*. She is smaller than a pinhead, but may produce as many as 500 eggs during her lifetime. At summer temperatures, the eggs can hatch in as little as 12 hours and the generations fan out rapidly if nothing kills them, 45×.

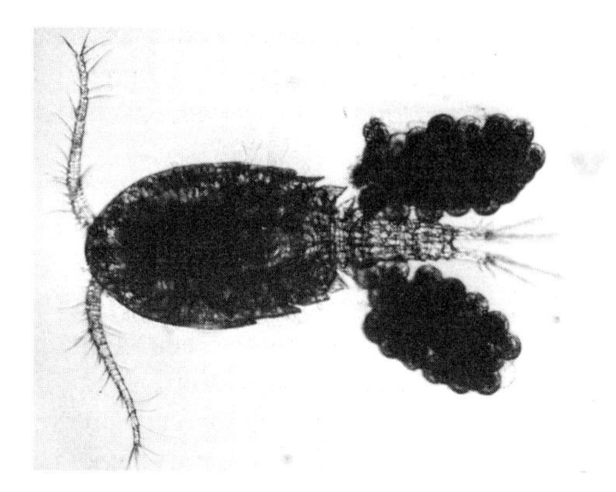

FIGURE 7-13. A female *Cyclops* with egg sacs, 45×.

The females exceed the males by a great margin. Some species reproduce all year round, but in spring and summer all are particularly fertile. It has been calculated that a single female would give rise to three million offspring in one year if there were no losses among the many generations.

The female starts producing eggs shortly after mating. The eggs emerge from two oviducts on the sides of the abdomen, and are fertilized as they pass the seminal receptacle in which spermatozoa have been deposited during copulation. A gelatinous matter is secreted to form a sac (see Figure 7-13) in which the eggs are held together. During her lifetime, a female may produce up to 13 pairs of ovisacs, each consisting of up to 50 eggs; usually the spermatozoa in the seminal receptacle are sufficient to fertilize all the eggs laid by the female during her lifetime.

To chase a *Cyclops* under the microscope is an exciting adventure. The animal, one of the "waterfleas," jumps around in a jerky manner, changing directions capriciously. But finally the one pictured in Figure 7-13 came to a stop long enough that even Daguerre could have photographed it. This was also an opportunity to observe the animal at higher magnification. And, lo and behold, some movement could be detected in the egg sacs! Indeed, after a while the first *Nauplius* hatched, a rare stroke of luck (see Figure 7-14).

The *Cyclops* undergoes a long and complicated body change. The newly hatched larva, called *Nauplius* (see Figure 7-15), looks very different from the adult animal and was long considered to be a different genus, although as early as 1699 van Leeuwenhoek suggested their relationship. The larvae undergo eleven and, in some species, twelve moltings. When the young animal hatches, it is oval and has three pairs of appendages, which later develop into two pairs of antennae and the mandibles. In the *Nauplius* stage, these appendages serve as swimming organs. After successive moltings, more appendages appear, which later enable this little animal to shoot through the water with surprising speed. The hind end develops, dividing the body into two parts; finally the adult stage is reached. The animal is then a streamlined, elegant creature, about 1 mm in length from eye to tail.

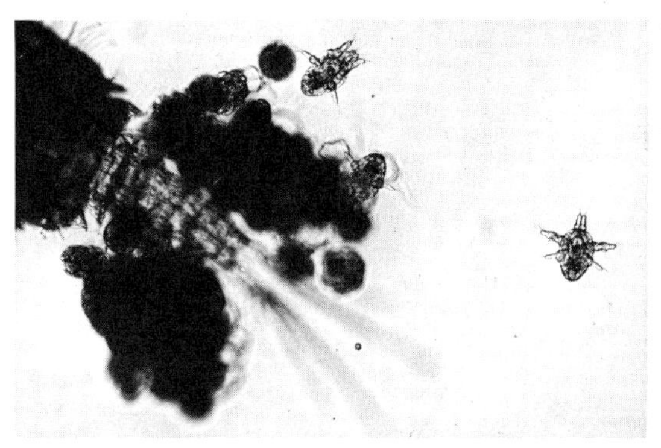

FIGURE 7-14. The same animal a few minutes later. Delivery is in full swing. One of her little offspring can be seen on the right side of the picture making its way cautiously over the slide. It "sees the light," the light of a 2000-candle-power microscope lamp, 50×.

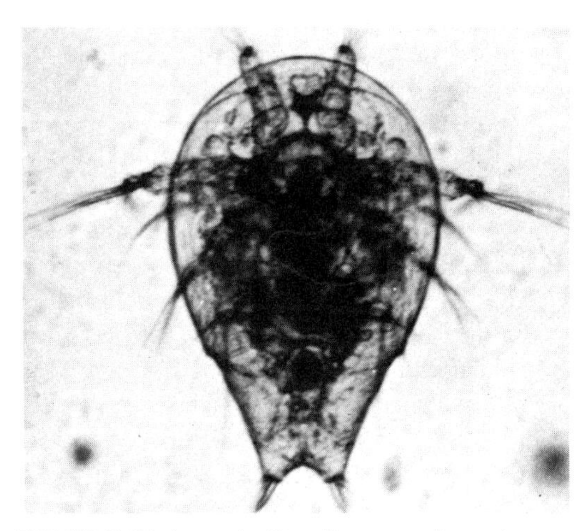

FIGURE 7-15. An early *Nauplius* stage from the same culture. The larvae go through 11 or 12 molts, 190×.

At high temperatures, the development from the fertilized egg to the moment of hatching takes as little as twelve hours; in winter, it can take as much as five days. The metamorphosis also depends greatly on temperature.

Planaria: The Amazing Worm

If you cut your finger, you usually look forward to its healing with confidence. Nature provides for a way to correct the damage. It is still not completely understood how this marvel works. Yet wound-healing in the higher

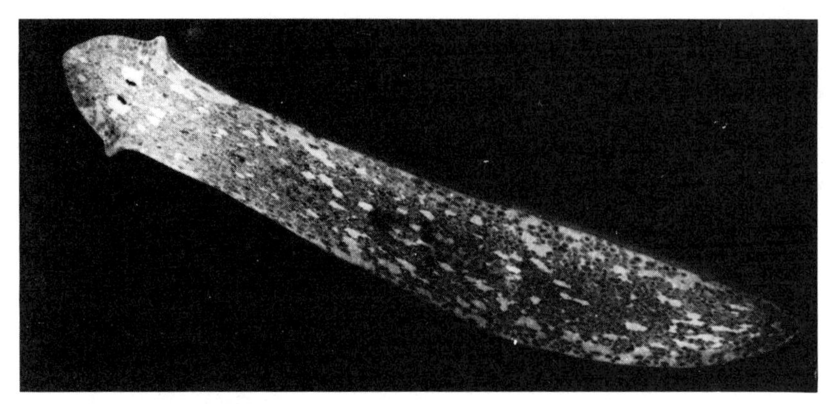

FIGURE 7-16. The Planarian *Dugesia tigrina,* 10× (photographed with reflected light).

animals cannot be compared with the feats some of the most primitive creatures are able to perform.

We can repair bodily harm, more or less perfectly. But many of the lower animals are able to replace entire parts of their bodies. Lizards, when they lose a leg, can regenerate it completely. Many fishes regenerate scales and fins. If a *Paramecium* or a *Stentor* is cut in half, both halves will develop in complete animalcules, provided each fragment retained part of the macronucleus. More than 200 years ago, the Swiss naturalist A. Trembley established his fame for all time by decapitating a *Hydra* and discovering that within a few days, the animal's body had grown a new set of tentacles while the disjoined headpiece developed a new body!

No less remarkable are some flatworms, among them *Planaria* (see Figure 7-16). *Planaria* can be found in brooks, streams, rivers, and lakes, where they crawl about under stones and decaying leaves in search of dead animals whose remains serve them as food. They are 10 to 15 mm long, with a flat body. *Planaria* boast a number of organs: a primitive brain, a nerve system, and two eyespots, which cannot perceive more than the light intensity of their surroundings. The animals shun the light, being what is called negative phototactic. Because of this sensitivity, they thrive best in the dark.

Planaria's reproductive apparatus is complicated for a "primitive" worm. *Planaria* are hermaphrodites—each individual has both male reproductive organs (testes, sperm duct, seminal receptacle) and female reproductive organs (ovaries, oviduct, uterus). Nevertheless, mating is necessary for fertilization.

When a worm reaches its optimal size, it breaks spontaneously into two parts. The head part develops a tail, the tail part a head, which brings us to the most remarkable feature of these animals: their ability to regenerate lost or injured parts of the body.

We know how the healing process takes place in humans, though not in all detail. During injury, blood exudes and coagulates, thus providing a temporary closing of the wound. Then epithelium cells—the cells of the epidermis—begin to divide and to migrate toward the center of the wound. At the same time, the division rate of the underlying connective tissue increases. We do not know what actually constitutes the stimulus that starts this process after the injury occurs, nor why the cell division stops when the wound is repaired. The essential point is that the healing mechanism in humans involves population growth of cells that already represent the needed cell type, having been "differentiated" beforehand.

Regeneration is quite a different matter. Figure 7-17 shows a worm that I cut into three parts. Two weeks later, I had three worms instead of one, all three complete and vital. The regenerates (see Figures 7-18 through 7-20) are smaller than the original specimen because the patients cannot feed during the recovery period. They will, howeer, reach normal size when properly fed.

Making an incision into the head is another problem-loaded operation that results in a double-headed monster (see Figure 7-21). Each head develops its own brain and two eyes. Looks the world brighter to the worm with four eyes? Is the worm in peace with itself, or are the two brains fighting each other, the left one insisting to make a left turn while the other one forcefully pulls to the right? The possibility of a personality split cannot be lightly dismissed.

The first observation of regeneration in flatworms is attributed to P. S. Pallas, an Italian, who reported it in a book published in 1774. Since then, scientists have grappled with regeneration without solving the mystery. This question of growth—both controlled and uncontrolled—lies at the center of the efforts of the embryologist, the geneticist, and the cancer researcher alike. The embryologist follows a controlled, predetermined development: the gradual change of a fertilized ovum, which finally results in

FIGURE 7-17. *Dugesia tigrina* before having been cut into three fragments, 9×.

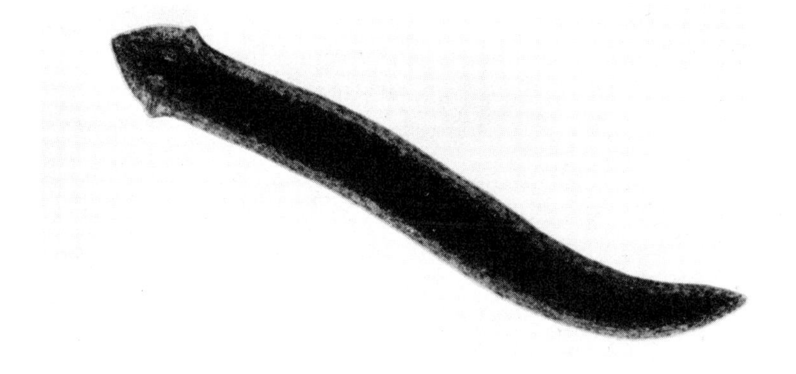

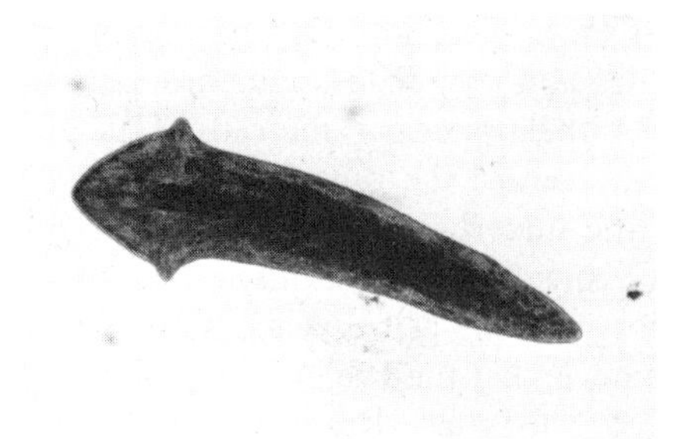

FIGURE 7-18. Regenerated head fragment, 9×.

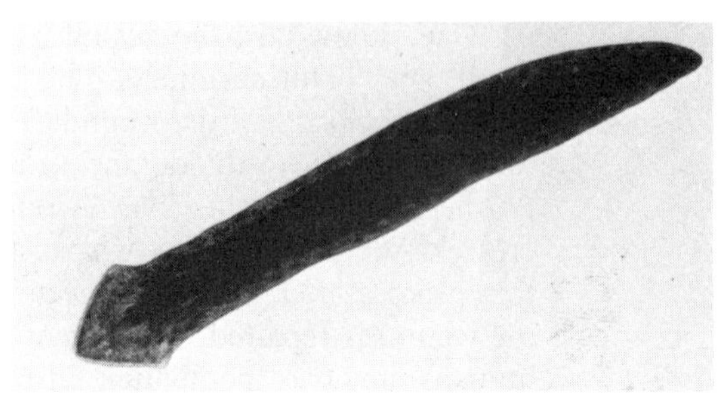

FIGURE 7-19. Regenerated middle fragment, 9×.

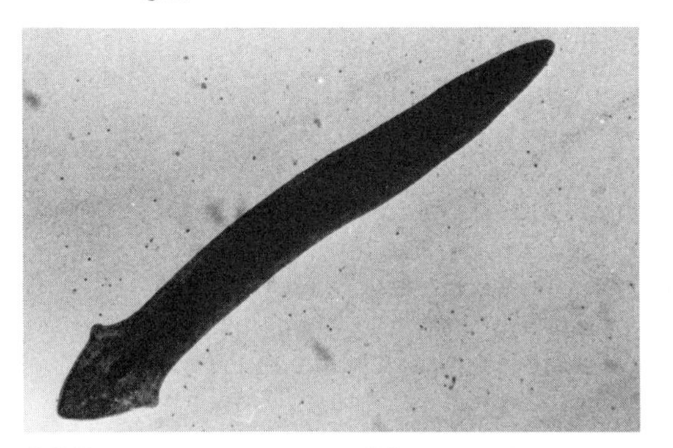

FIGURE 7-20. Regenerated tail fragment, 9×.

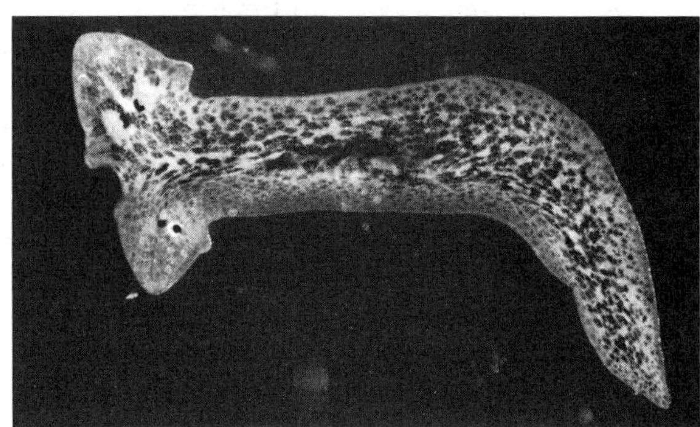

FIGURE 7-21. Two-headed monster, 8×.

a complex organism. The cancer researcher, on the other hand, is interested in the cause of uncontrolled, anarchic growth—a tumor.

In flatworms, regeneration can be seen as the resumption of an interrupted embryonic development. According to the current explanation, a flatworm's body is interspersed with thousands of embryonic cells, called neoblasts, each of which has the potential to develop into any cell type, or to "differentiate." Neoblasts (see Figure 7-22) are very small cells, about 10 to 12 microns in size (one micron is 1/1000 of a millimeter). They have a large nucleus, which fills almost the whole cell, leaving only limited space for the cytoplasm. There are usually three nucleoli, though the number is not always certain because it is difficult to distinguish them from the granular material in the nucleus. A fully grown worm has thousands of these cells (in one species, 15 mm long, the total number was estimated at 37,820 plus/minus 5 percent).

A neoblast can become a brain cell or an eye cell, or it can form part of the digestive tract or reproductive apparatus. When a worm is cut in half, the neoblasts migrate to the wounds and accumulate in a structure called the blastoma. Differentiation now sets in. The central neoblasts of the tail fragment first forms the brain; the next group organizes the eyes. Other or-

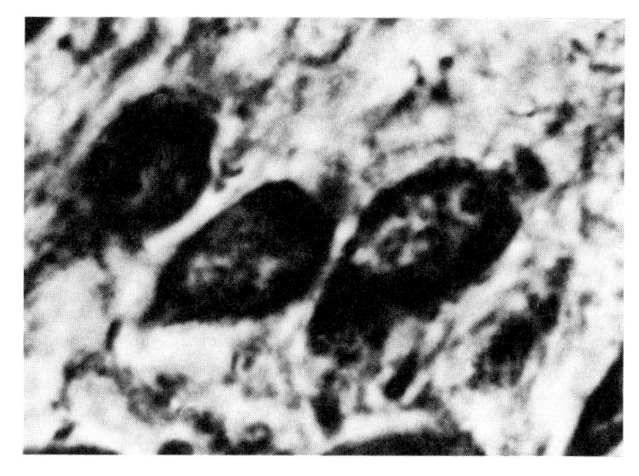

FIGURE 7-22. Neoblasts, 2.500 ×.

gans, muscles, nerves, cilia, and so on, follow until the fragment is once again an individual capable of all vital functions. The wonderful thing about this process is that only the right number of neoblasts form the brain, eyes, and so on. Then a mysterious mechanism operates to inhibit the process and prevent overgrowth. The neoblasts that are not needed remain undifferentiated. They persist as a reserve for possible future injuries or for asexual reproduction through fission. It is this mechanism of inhibition that interests the cancer researcher.

Another question is why the head fragment regenerates a new tail and not a second head. How do the neoblasts receive the proper information? If we had the answers to such questions, we would be able to control any healing process.

The Vinegar Eel

A microscopic animal that brings forth living young is the vinegar eel *(Anguillula aceti,* also called *Turbatrix aceti)* shown in Figure 7-23. In *Anguillula,* "childbirth" goes on continuously. A single female was once observed to give birth to 45 young in her lifetime. In some pregnant females, several embryos can be seen lined up as on an assembly line. Sometimes these demonstrate all stages of embryonic development, the most mature near the end of the worm, the least advanced near the head. Figure 7-24 shows three such embryos: the one on the right is almost ready for "delivery"; beside it is a younger one, its wormlike nature already evident; next to this is part of a third "baby," less advanced in its development. They all occupy a long, outstretched uterus. A newly born worm is about 1.5 mm long. Sometimes the mother dies before the young are ready to hatch. If this happens, the later stages keep on growing and can be observed moving about in their mother's dead body.

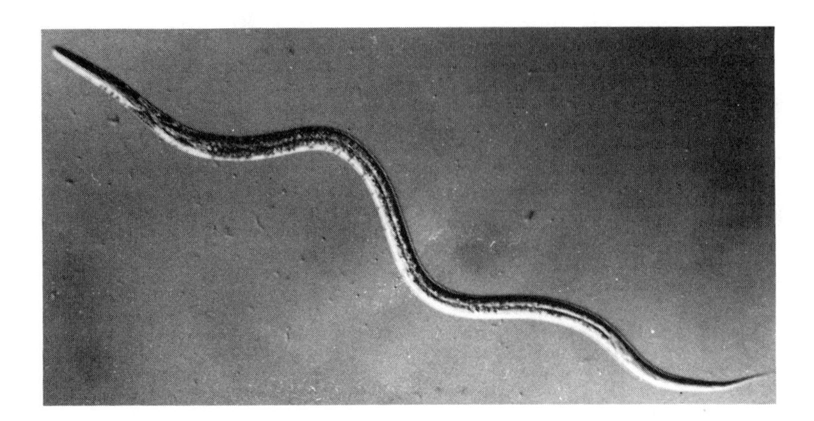

FIGURE 7-24. Vinegar eel with embryos in utero, 410×.

Vinegar eels are so small that we had to wait for the development of the microscope to make their acquaintance, even though they probably sneaked into Noah's vinegar bottle in order to survive the flood. Pierre Borel, the private physician of Louis XIV and author of one of the earliest books on the wonders of the microworld,—*Observationum Microscopicarum Centuria* (1656),—first noticed the eel while raiding his wife's kitchen. For hundreds of years, our ancestors had enjoyed salads prepared with vinegar, relished cheese together with its cheese mites, sipped water from ponds and streams. Now it turned out that all these things were teeming with horrible creatures! As Borel describes it:

> *In vinegar we can see with the help of the microscope snakes, or eels, or rather snakelike worms which move with great speed. They try to get to the surface of the liquid in order to breathe. Many people to whom I have shown them abstain now from taking vinegar. It is amazing that living creatures can exist in such a sour substance. Yet perhaps we should not be too astonished about it if we realize that living beings which resemble whales, hide in hot springs and in our own body, yes in our blood.*

Borel's contemporaries were horrified by these disclosures. Then the popular belief gained ground that the acid taste of vinegar was due to none other than the vinegar eels, which strike our tongues and palates with their pointed tales. People did not want to forego their salads—an excuse had to be found to show that these worms were "necessary."

Today, we vainly search our vinegar bottles for eels. Not that they are extinct; they still show up during the manufacture of vinegar. But in the finished product, they are killed by small quantities of sulphuric acid or by pasteurization and eliminated through filtration. Yet there was a time when these "snakelike worms" served as popular objects for the microscope and played an important role at social gatherings, when the microscope lamp was kindled after supper as we might turn on our television set today in order to entertain our guests. These animalcules were also fraudulently used by quacks. The patient was invited to look at his "natural juices" through the microscope and induced to believe that the teeming mass of worms originated directly from his or her body. Naturally, the patient then gladly submitted to any costly treatment the quack suggested.

For a long time, scientific periodicals discussed whether the ingestion of vinegar eels was injurious to health. An English zoologist, B. G. Peters, settled this question in the 1920s, once and for all, in a heroic way—twice, at two-day intervals, he swallowed luxurious cultures of vinegar eels, approximately 36,000 individuals each time. The result was that the swallowed worms were completely digested without leaving a trace.

There remains one problem: where do vinegar eels come from? Vinegar, after all, is a manufactured product.

A natural source exists for acetic acid. It is found in certain slimy discharges of trees, which are caused by injuries to the bark. A number of microorganisms have been observed in these discharges, among them nematodes, which resemble vinegar eels. These discharges undergo fermentation, during which first alcohol, and then acetic acid is produced. It is possible that these discharges constituted the original habitat of vinegar eels until they were transferred to the manufacturers' vats by flies, where they have remained ever since.

Vinegar eels are available from biological supply companies and are very easy to culture. They must be ordered with the medium, which consists of unpasteurized cider with "mother of vinegar" added. On arrival, the culture is transferred to the medium, some of which can be saved for later subculturing. I have also done subculturing with apple cider vinegar from the grocery store; it generally works well, depending on how the vinegar was treated by the manufacturer. But it is advisable to keep the old culture for a while to make sure that the subculture "takes."

Though easy to maintain, these worms are not so easy to observe alive, due to their remarkable, tireless agility. The observer must wait and keep a sharp eye on the vinegar drop under the microscope until the animals are immobilized by the coverglass, when the fluid gradually evaporates. Holding the slide over a lighted match for three to four seconds kills the animals without distorting them. The photomicrographs were taken from the living specimen with electronic flash.

8
Of War, Peace, and How to Hitch a Ride

Warfare is a natural activity throughout the living world, including the microcosm. Some animals, destined to be carnivores, must prey on other animals in order to survive or to provide food for their offspring. But many other dwellers of the invisible world have mastered the difficult art of peaceful coexistence. These are the symbiotic relationships that we find among plants, between animals and plants, or between different animals. Both activities offer remarkable ways of getting along in the microworld.

Warfare in the Microworld

We humans are somewhat overimpressed by our own resourcefulness. It is true that our ancestors probably could not have survived without well-developed brains for devising substitute weapons; but we often overlook the fact that their spears, knives, and poison arrows were preceded by similar devices used by other creatures since the dawn of life. Many of the more ingenious of these weapons are too small to be seen with the unaided eye. Yet some of them would greatly tax humanity's inventiveness if we were to try to copy them, even today.

The deceptively sluggish *Hydra,* for instance, kills with lazy efficiency by lashing out poisonous whiplike threads at its victims. It remains almost stationary on a rock or other hard surface, with tentacles (see Figure 8-1) spread and moves them spasmodically now and then. The tentacles are

FIGURE 8-1. One of *Hydra's* tentacles at high magnification, showing the treacherous nettle capsules, 200×.

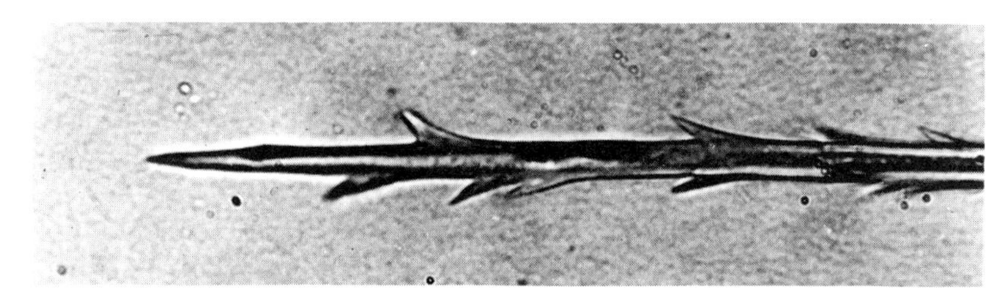

FIGURE 8-2. The poisonous hair of a European processionary caterpillar, 1025×.

studded with bodies called nettling capsules, which are comparable to tiny guns. Each of these contains a poison in a hollow, barbed thread coiled like a hairspring. When a potential victim, perhaps a waterflea, brushes against *Hydra*'s tentacles, some of the capsules explode and eject their threads. These strike the prey and paralyze it with poison from their tips. *Hydra* is then ready for a leisurely meal.

Hydra's guns are more efficient than a human's. They will not fire accidentally, but only when organisms suitable for food touch the tentacles. Amazingly enough, they will not discharge when touched by a leaf or a stone, or even when irritated by a parasitic polyp-louse crawling along their surfaces. After use, *Hydra* develops new armament within a few hours.

The larvae of certain European moths have microscopic weapons capable of inflicting injury, even upon humans. These larvae are the processionary caterpillars, so named because of their habit of moving about in meandering armies up to six feet long. This habit makes them extremely vulnerable to birds, so the poisonous hairs (see Figure 8-2) they have evolved should be classified as defensive weapons—hairs so brittle that they are sometimes broken off and carried by the wind into the eyes and mouths of unsuspecting hikers, causing acute inflammation. Forests in some parts of Europe have been closed to the public when dangerously infested by the processionary caterpillar.

Much of the fighting in the microworld is chemical warfare, its daggers and spears being used for injecting poisons. The solitary wasp has refined this technique for the purpose of furnishing "strictly fresh" food for its larvae offspring. Wasp poison is so constituted that the weevils, caterpillars, or flies into which wasps inject it are not killed, but indefinitely anesthetized, to become banquets for future wasp families.

A dramatic instance of this natural anesthesia is the performance of the wasp *Cerceris tuberculata* Klg., which exclusively chases a hapless weevil called *Cleonus opthalmicus*. When *Cerceris* locates *Cleonus*, she grabs it and, during a brief struggle, plunges her stinger into a nerve center in its chest. *Cleonus* drops, and the wasp lifts it and carries it off to her nest. She repeats such conquests until, several weevils later, she is ready to deposit her eggs and seal the nest, stocked with fresh food for her brood.

Another wasp, *Eumenes pomiformus* Fbr., catches caterpillars for her offspring, but apparently is unable to paralyze them completely; after anesthesia, they still move when touched. *Eumenes* makes up for this imperfection by suspending her eggs from threads attached to the ceiling of the nest. As the larvae hatch and begin to feed, they can escape injury by climbing up the threads when the caterpillar begins to struggle.

The venom that accompanies a wasp's sting (see Figure 8-3) is a mixture of two chemicals produced in separate glands. One gland secretes acid into a large poison sac; the other secretes an alkaline fluid into the base of the sting. The fluids are mixed when the wasp uses her stinger, and the mixture is more effective then either would be alone. Two fingerlike sense organs enable the wasp to pick the best spot for injection; strong muscles pump in the poison, and two more glands are believed to lubricate the various parts involved—all in all, a very complex war machine.

A worker bee's stinging apparatus (see Figure 8-4) is much the same, except that it is ordinarily used only once, while the wasp's may be used

FIGURE 8-3. The sting of a wasp, detached from the body.

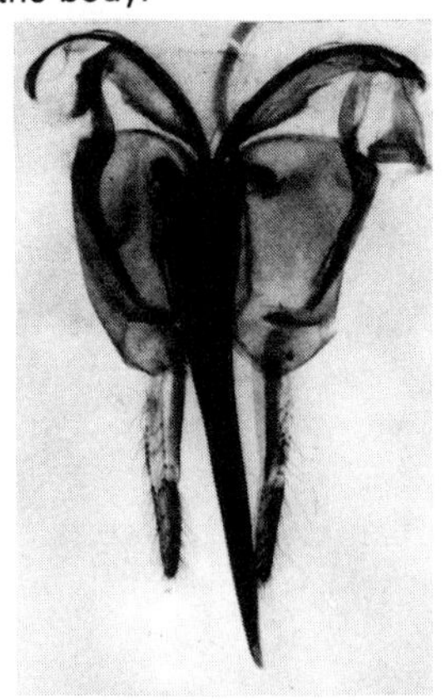

FIGURE 8-4. Barbs on the sting of a worker bee hamper its removal after use. The bee dies as the sting is wrenched from its body, 100×.

FIGURE 8-5. Miniature syringes of the stinging nettle (15×) are sealed by tiny knobs at their tips (enlarged view, 530×, below). In action the knobs break off, the needle pierces the skin, and the poison from their bulblike bases squirts into the wound.

many times. The bee's sting is barbed, and the bee has great difficulty extracting it after it has attacked the intruder. So a bee usually suffers fatal injuries performing a miniature kamikaze. The complete stinging apparatus, with poison sac attached, is wrenched from the body and left sticking in the wound. The detached muscles keep on working to drive the sting in deeper and deeper, while more and more poison is pumped into the tissue.

All of these tiny weapons inevitably recall familiar instruments in the doctor's office. In the case of the European stinging nettle, the similarity is almost uncanny.

The stinging nettle is equipped with many thousands of stiff hairs, constructed like tiny syringes (see Figure 8-5). A vesicle at the base of each hair contains a poisonous fluid, mainly formic acid, and the hair itself is actually a long, hollow, very brittle needle, closed at the point by a tiny knob. At the slightest touch, the knob breaks off, the hair penetrates the skin, and the resultant pressure against the base forces toxic fluid into the wound. It is a fine weapon against herbivorous animals, though not good enough to withstand occasional attacks by "old country" housewives, who—armed with gloves—harvest nettles as spinach when other food is scarce.

The microweapons with which humans are most acutely familiar are the tubelike mouthparts of bloodsucking parasites, such as mosquitoes (see Figure 8-6), bedbugs (see Figure 8-7), fleas, and lice. One of their weaknesses, fortunately, is that while thrusting their needles, stylets, or mandibles into our skin, the insects must simultaneously inject saliva into the

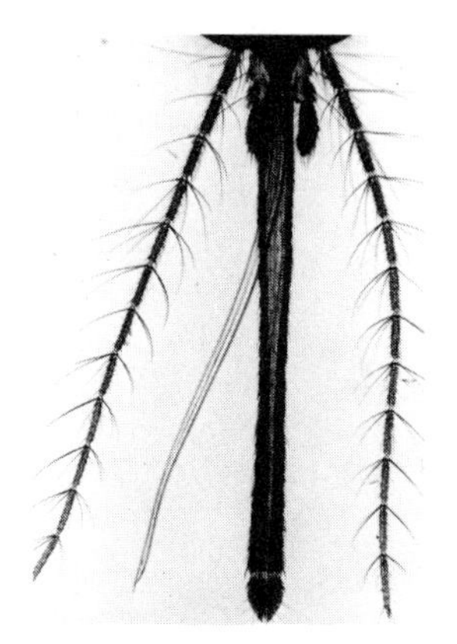

FIGURE 8-6. Proboscis of a female mosquito. The sucking tube is separated here from the heavier sheath. The male is unarmed and subsists on plant juices instead of blood, 28×.

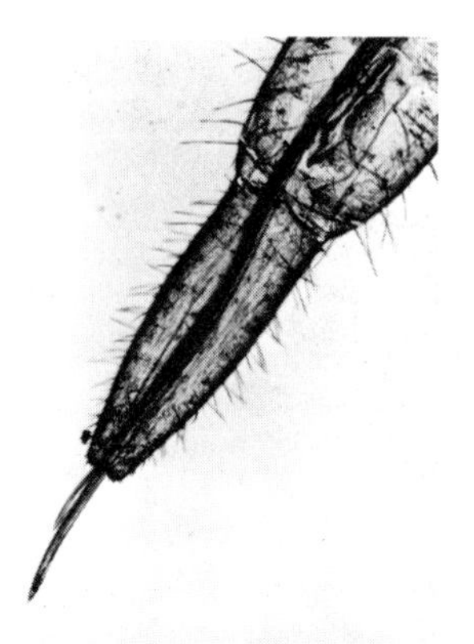

FIGURE 8-7. The bedbug's proboscis has tiny stylets that pierce the skin and allow the blood to flow, 85×.

wound to keep the blood from clotting. This creates a momentary irritation that warns us that we are being attacked and spurs us to counterattack.

A bedbug needs five to ten minutes to get a satisfying meal. Furthermore, it requires at least one hearty meal between moltings in order to develop into a healthy, robust adult. This is a full-time job when you consider that the bedbug molts about three times before maturing.

A few hours of peering through the microscope at the fierce world of the bedbug and its confreres makes one realize with a shudder that—armed with their knives, stilettos, needles, tentacles, poisons, and anesthetics—all these creatures might have carved quite a different niche in the scheme of life if nature had made them larger.

Life With an Alga

There are many examples in nature to show how totally different organisms can live together in such close association that their coexistence is equally vital to both partners. In these relationships, the microworld is frequently represented. Bacteria, protozoa, and microscopic fungi are all found as partners united with countless, very different animals and plants. The most remarkable group of such symbiotic "comrades" are certain microscopic algae, which have established many relationships with other organisms, large and small.

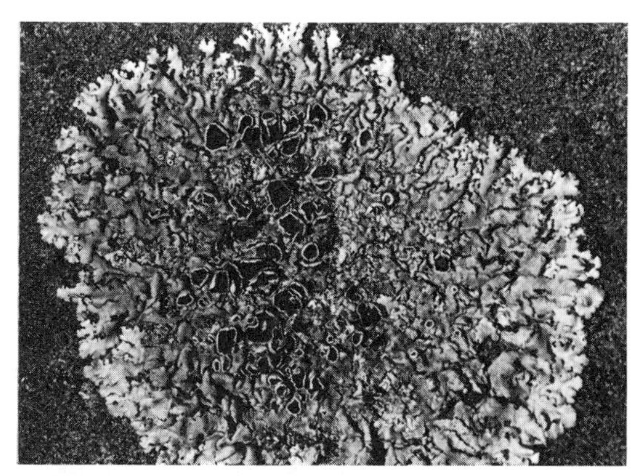

FIGURE 8-8. *Parmela rudecta,* a lichen of gray-green color that grows on barks and rocks.

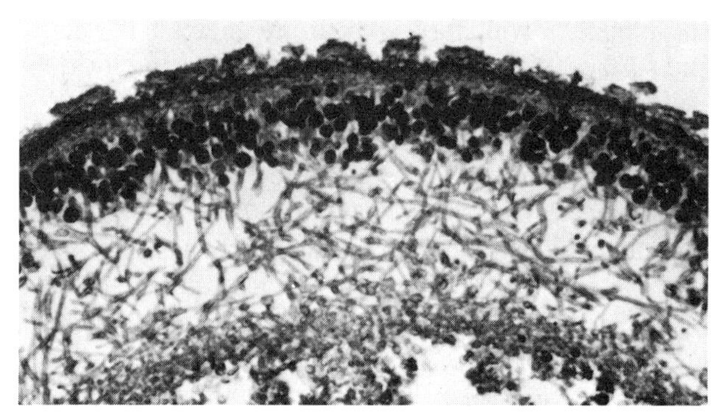

FIGURE 8-9. Cross section through a *Parmela* lichen at higher magnification. The tiny round algal cells are entangled in the mycelium of the fungus, 110×.

One of the most common and most successful of such cooperatives is found in the lichens, met with almost everywhere (see Figure 8-8). They grow on stones, rocks, trees, and walls, often clinging to substrates on which no other plant can take hold. In some parts of the earth, they are the only plants capable of surviving the most adverse environment, such as the extreme temperatures of the polar regions. They consist, in fact, of two very different plants, a fungus and an alga (see Figure 8-9), which maintain a kind of common housekeeping. The fungus belongs to a group of plants lacking in chlorophyll, and is unable to produce the carbohydrates a green plant synthesizes from carbon dioxide and water under the influence of sunlight. For the fungus, it is of great importance to harbor among its fine hyphae a minute alga, which has just this ability and is willing to share whatever it produces in excess of its own needs. The alga deposits microscopic starch granules on its cell wall, which the fungus readily absorbs. As a reward, the alga not only finds shelter and protection, but also profits from the ability of the fungus to take up water and store it, a contribution which, in turn, is life-saving for the alga. The fungus also provides certain inorganic substances the alga needs. By secreting acids and enzymes, in an infinitely slow process, the fungus dissolves the substrate on which it grows and uses it for food.

The mutual dependence between these two plants is so great that neither can reproduce without the other. The fungus that, as a rule, makes up the bulk of the lichen, produces fruiting bodies whose spores are dispersed by the wind. These spores are unable to germinate unless they find, by chance, a few algal cells of the proper kind with which to unite. In this case, the alga reproduces by fission, while the fungus develops a new mycelium; a new lichen takes hold and grows. In some species of lichens, fungus and alga travel together to a new home. These lichens form what are called soredia—small clumps of hyphae in which a few algae are entangled. The soredia break through the surface of the lichen and are then carried away by the wind.

Until recent times lichens were believed to be the most perfect example of a symbiotic relationship in nature. After all, they were so successful that about 12,000 different species of lichens thrive today on the earth.

However, this notion of a symbiosis par excellence has now been questioned. Research of the last two decades has revealed that the mycobiont in some lichens parasitizes its phycobiont. The foremost proponent of this theory is the botanist Vernon Ahmadjian. He has found evidence that the hyphae in some lichens develop specialized outgrowths, called haustoria, that invade the algae to feed on their cytoplasm, thus killing them. It even has been observed that in a single lichen plant the fungus may parasitize only algae in one area of the plant, sparing those in a neighboring area. The truth about the nature of lichens, therefore, lies in between, as Ahmadjian himself suggests when he says: "To say that all lichens are mutualistic is as wrong as saying that all are parasitic."[1]

Nobody knows how it happened that the three-toed sloth *(Bradypus tridactylus)* first found an alga in its fur. But the outer hairs of its pelt changed in a strange way: they developed numerous deep crevices in which algae apparently find especially favorable living conditions (see Figure 8-10). For the sloth, this is literally a matter of life and death. Buffon, the famous eighteenth century French naturalist, said of this peculiar animal: "One more mistake and it would not have survived." Indeed, there is hardly another creature less well equipped for life. With its phlegmatic disposition, it is unable to escape quickly if danger threatens: Its safety therefore depends to a great extent on its ability to adapt to its environment. And this adaptation is precisely what the alga provides as a reward for the sloth's hospitality. There are, in fact, two different algae, a greenish and a reddish

FIGURE 8-10. The outer hairs of the three-toed sloth's fur have deep crevices in which two different algae live, 60×.

one. Together, they give the sloth's fur just the right hue to permit the animal to continue its lazy life in South America's jungles.

Closer to home, if a drop of pond water is put under the microscope, the green slipperanimalcule, *Paramecium bursaria,* whose nuptials we have already witnessed (Chapter 7), is sometimes encountered. It owes its color to the *Chlorella* alga. Although this is one of the most common algae, quite capable of leading an independent life, it willingly accepts the hospitality of this species of *Paramecium* and no other. And it is as cordially welcomed (see Figure 7-4).

Chlorella's benefit is evident: As a green plant, it needs carbon dioxide, a basic component for its photosynthetic activities. Carbon dioxide is given off during the animal's metabolism, so the plant finds a welcome source for an important chemical inside the animal. The alga is also well housed and protected.

Paramecium's benefit is not so apparent, and some observers feel that this relationship is one-sided, in that the animal serves the plant as a host without being harmed, but also without receiving any benefit. Yet there are indications that the two organisms are interwoven in a much more intimate way, and that the animal also depends on the alga to a certain degree.

If *Paramecium bursaria* is put into a medium that does not contain any organic food—such as bacteria, its normal diet—it does not starve, but continues to thrive and to reproduce, provided favorable living conditions for the alga are maintained. This shows that the animal can get from the plant all the nourishment necessary to keep it alive.

Another easy experiment consists of putting a culture of *Paramecium bursaria* into total darkness. After about two to three weeks, significant changes begin to take place. By depriving the animals of light, *Chlorella's* photosynthesis is disrupted. Taking the culture out of the dark, it is clear that the number of green cells in each animal has declined considerably. Small crystals have developed inside the animals. These crystals can easily be made visible with polarized light and appear as small, shiny particles (see Figure 8-11). Their nature is not known, though many suggestions as to their chemical composition have been made. The same crystals occur in other protozoa under certain conditions and have been identified as uric

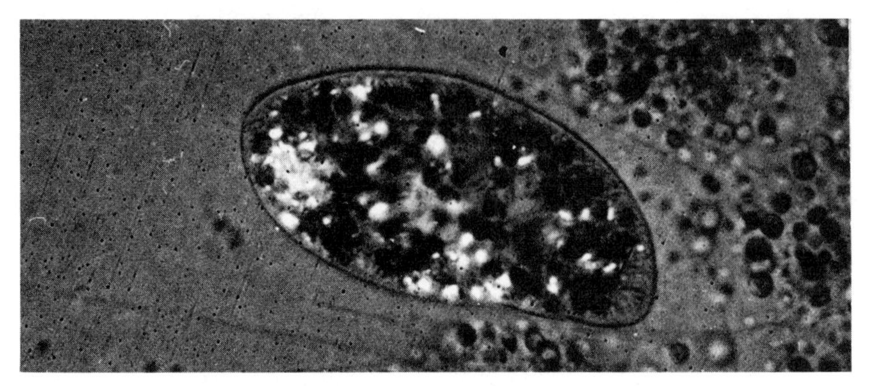

FIGURE 8-11. *Paramecium bursaria* after having been kept in total darkness for two weeks. Crystals have formed inside the animal's body that indicate basic changes in its metabolism. The algae have greatly decreased in numbers, 500×.

acid, calcium oxalate, calcium carbonate, and other compounds. In *Paramecium bursaria,* they are considered to be waste products, indicating a change in the animal's living conditions. If the algae are no longer capable of offering the animal sufficient food, the *Paramecia* depend more and more on the bacteria in the water, a change of diet that has a profound influence on their metabolism. The appearance of crystals indicates an accumulation of crystallized waste products.

It is remarkable that none of the other many species of *Paramecium* has adopted an alga. *Chlorella,* however, has found many other partners, among them the green freshwater polyp *Hydra viridis,* with which the alga lives under similar conditions.

There is one microscopic worm which, in its sad affair with an alga, sounds all too human. The worm is known as *Convoluta roscoffensis*; the alga is a species of *Chlamydomonas.*

Convoluta is a worm about 3 mm long (see Figure 8-12). It is a spinach-green color—here again, the presence of an alga gives it this color. *Convoluta* occurs only on the wide beaches of Normandy and Brittany, in France—nowhere else in the world—but in such masses that the wet sections of the beach on sunny days and at ebb tide are covered with green patches, huge colonies of sun-bathing worms.[2]

Convoluta is a primitive worm. It has no heart, no blood circulation. Two eyespots make it sensitive to light; these are important organs that tell the animal when the sun is shining, so that it can leave its hiding place in the sand to expose the green-celled algae to the sun's light, which is vital to the algae's well-being. Between the two eyespots is a small cavity called an otocyst. This cavity, lined with nerve tissue, contains a small piece of chalk (see Figure 8-13). The mechanism functions as an ingenious alarm system: The chalk grain lies loose in the cavity, and when the waves of the

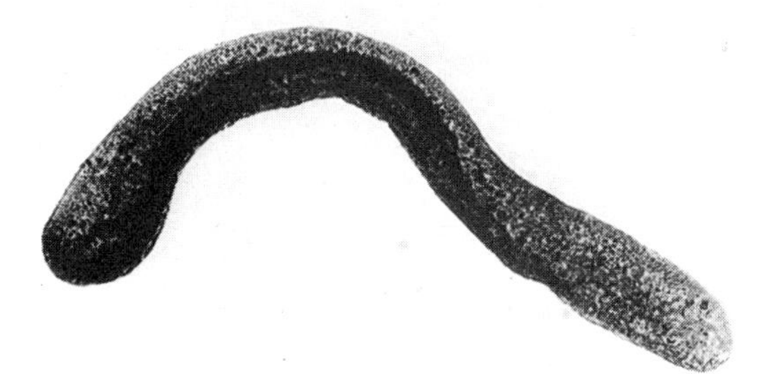

FIGURE 8-12. *Convoluta roscoffensis.* The algae with which this worm lives accumulates at the posterior end of the animal; therefore the head appears of lighter shade, 68×.

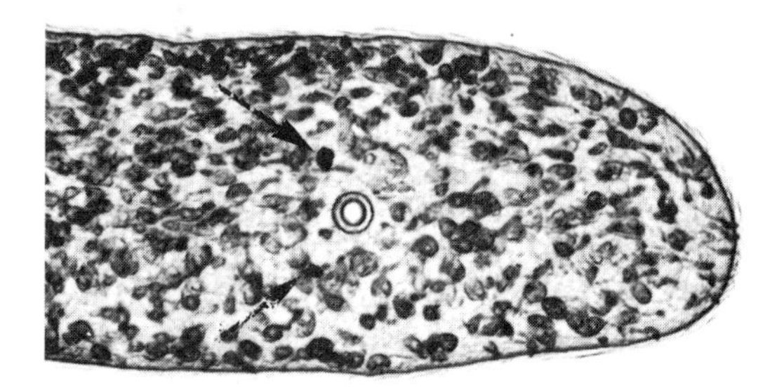

FIGURE 8-13. Head of *Convoluta* at higher magnification. Shown are the eyespots (arrows) and the otocyst in the center. Some algal cells are dispersed in the tissue, 240×.

sea pound the beach, the grain vibrates. This vibration activates the sensory cells of the nerve tissue, thus warning the animal and enabling it to escape damage from the approaching waves by hiding in the sand.

Compared with this highly developed and sophisticated organ, *Convoluta's* digestive apparatus is very primitive indeed. It has a mouth opening, but no digestive tract and no anus. The food is collected in a mouth cavity and distributed at random to different parts of the body. Digestion takes place in food vacuoles that form wherever food collects. They disappear when the digestion is completed. Whatever cannot be digested is eliminated through the skin, which breaks, pushes out the waste, and closes again.

Particularly remarkable is the fact that *Convoluta* lacks any system that would free it from the nitrogen products of its metabolism. In other words, *Convoluta* has no kidneys. The waste remains in its tissue. This a serious defect, which could well have deadly consequences, had not *Convoluta* solved the problem, as we shall see.

Convoluta is a hermaphrodite. Each individual has male as well as female sex organs. The eggs are fertilized inside the worm's body and laid in the sand. While being deposited, a gelatinous substance is secreted and forms a capsule that holds five to fifteen eggs together (see Figure 8-14).

When the young creature hatches, it is colorless. It behaves like all larvae, feeding on anything that can serve as food. But after a few days, the worm gradually becomes green. A week later, something very strange happens: *Convoluta* stops feeding. From now until it dies, under natural conditions, it will never again ingest solid food. The reason is that it is content simply to absorb the carbohydrates that the green cells in its body produce. As soon as the green cells have multiplied sufficiently, the worm gives up making any effort to search for food and relies upon the alga. Not only this; it allows the alga to fulfill another important task: to rid it of the waste products of its metabolism. The alga, in turn, needs the nitrogen for its own well-being. In this stage of their lives, the relationship between animal and

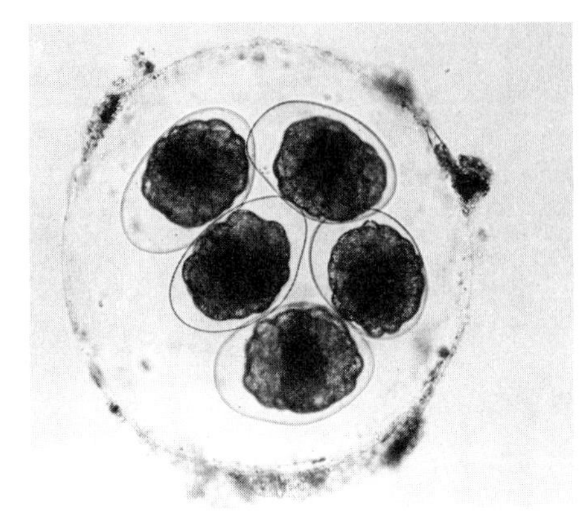

FIGURE 8-14. *Convoluta*'s egg capsule with five eggs containing well-developed embryos, 52×.

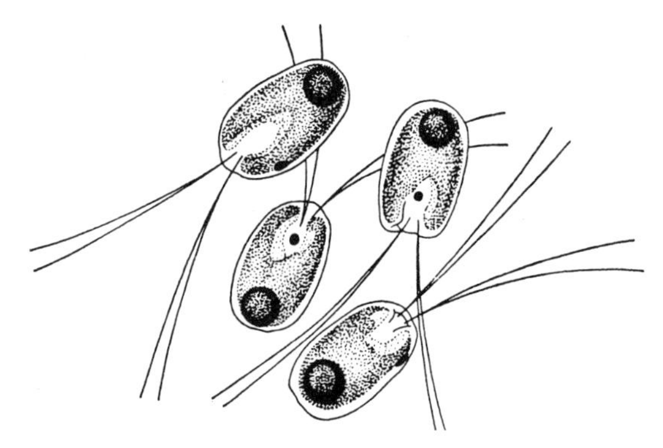

FIGURE 8-15. Group of *Chlamydomonas* algae (the larger form in the free-living stage), 73×. (*From* F. Keeble and F.W. Gamble, "The Origin and Nature of the Green Cells of *Convoluta roscoffensis,*" *Quarterly Journal of Microscopical Science,* Part 2, Vol. 51, (1907). *Courtesy Journal of Cell Science* as the successor to the *Q.J.M.S.*)

plant is a genuine symbiosis—both derive equal benefit from the association. This relationship lasts until the worm reproduces. But then another strange thing happens: *Convoluta* starts digesting its algae! As F. Keeble says, it bites off the hand that feeds it. The animal is driven to this amazing act by sheer hunger—in the long run, the green cells are not capable of sufficiently feeding the worm. But the worm also pays a high price for this action. Unable to return to a normal mode of feeding, the worm starves as soon as the algae are digested.

Chlamydomonas itself is a many-faceted organism (see Figure 8-15). Sometimes it is not easy to say whether it is an animal or a plant. When free-living—that is, when not together with *Convoluta*—the alga consists of an oval cell. It has a red eyespot and four flagella, which are attached to the anterior part of the body and enable it to move about like an animal. The cell is sometimes green, sometimes colorless. If green, it lives like a plant, manufacturing its own food; if colorless, it lives like an animal, taking up organic matter from the outside. To make things even more complex, this alga occurs in two sizes; the larger measures about 16 microns, approximately double the size of the small one.

After a time of lively activity, the flagellated cell comes to rest. It throws off its flagella, surrounds itself with a cell wall, and divides into four daughter cells. These cells develop their own flagellae and start a new, independent life.

But somehow, in a puzzling way, these active, free-living algal cells are irresistibly attracted by *Convoluta*'s egg capsules. Whenever they meet the capsules in their restless travels in the water, they stick to them or enter them by piercing the soft gelatinous cover of the capsules. They then throw off their flagella and start to divide. When the *Convoluta* larvae are ready to hatch, the capsule breaks open and countless unflagellated algae are liberated. This is the moment the young worm is inoculated with the alga—a moment of decisive importance for its survival. The infection takes place in a simple way: The alga enters the worm's body during the short period of time in which the animal ingests its food. The swallowed alga—a single cell is sufficient—remains undigested. Not only this; it starts dividing and reproduces until the space beneath the worm's surface is beset with green cells. But the process that takes place inside the worm is not a reproductive one. No eyespot or flagella develop. The cell wall is reduced to a thin membrane. Even more serious, the nucleus degenerates during the repeated cell divisions until not the slightest trace of a nucleus can be perceived in the green cells. The alga loses its ability to reproduce in this association. This is why it cannot be passed on to the offspring by the mother worm, so that each larva must be inoculated anew. The only function the algae cells retain undiminished is the ability to synthesize carbohydrates for the animal's benefit and to free it from the waste products of its metabolism.

How could such a bizarre relationship between animal and plant have developed? Keeble explains it as nitrogen hunger on the alga's part. Nitrogen is scarce in the sea, and the plant is seduced by the traces of nitrogen in *Convoluta*'s egg capsules.

Life in the Hindguts of Wood-Feeding Insects

There can hardly be a space in the entire world where so many different protozoa live as densely compressed together as they do in the hindgut of wood-feeding termites and roaches. It is over 100 years since early observers of termites first announced the presence of "parasites" in the insects' intestines; yet knowledge in this field is still incomplete. There are not less than 400 species of wood-feeding termites, and each has its characteristic intestinal fauna. Only about 50 species have been investigated so far for their protozoa, among them *Reticulitermes flavipes,* the species at home in the eastern United States. The huge family of the *Termitidae,* which alone embraces 1100 species, is disregarded in this estimate. These termites feed mainly on humus and plants, not cellulose; they therefore harbor few or no protozoa.

The hindguts of termites also teem with a rich flora, including spirochaetes, bacteria, and microscopic fungi. The intestines of these insects are universes in themselves and probably not peaceful ones. For the termite, some of these organisms are symbionts. These organisms are capable of transforming cellulose into glucose, which the termite cannot do because its digestive juices lack the enzyme necessary to break down the cellulose. The protozoa use part of the glucose themselves and allow the insect to use the excess. The termite also chews up the wood. In its midgut, the wood particles are further ground up into bits small enough for the protozoa to ingest and to digest. The flagellates furthermore find shelter and protection as a reward for their contribution in this remarkable symbiotic relationship.

Not all intestinal inhabitants of the termite are symbionts, though. Some live inside the host without harming, but also without benefiting it. The spirochaetes (see Figure 8-16), bacteria, and fungi parasitize some of the flagellates and serve others as additional food. Little is known about the interaction of these creatures, which may lead a bitter struggle for survival in this restricted environment.

Occasionally, termites are seen that do not contain any protozoa; such termites are about to molt. During this time, they undergo a period of fasting for about four weeks. They have no need of their symbionts because they do not eat wood at this time, being fed instead by termite workers with a special saliva. After molting, the termite regains its protozoan fauna quickly by feeding on the excrements of its co-citizens. The protozoa taken up in this way undergo a sudden population explosion and quickly establish themselves again in the termite's intestines.

Figure 8-17 shows the best-known protozoan from termite intestines, *Trichonympha gracilis,* one of the most common in wood-feeding insects. It lives not only in the termite of the eastern United States, but also in other species. The American physician and naturalist, Joseph Leidy, named it Trichonympha because the arrangement of the long flagella clothing the body, reminded him of "the nymphs in a recent spectacular drama, in which they appear with their nakedness barely concealed by long cords suspended from the shoulders" (*trichos* is Greek for hair).

Other common species are *Pyrsonympha vertens* and *Pyrsonympha major* (see Figures 8-18 and 8-19). Alive, the former is especially beautiful. Its four to eight undulating membranes vibrate constantly, giving the illusion of a burning torch (*pyrsos* is Greek for flame).

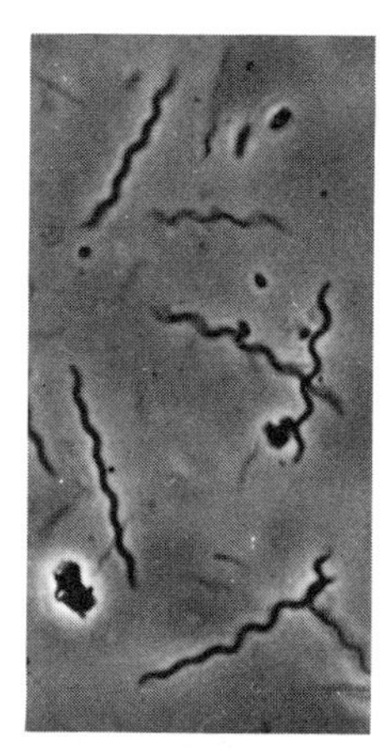

FIGURE 8-16. *Spirochaetes,* 450×.

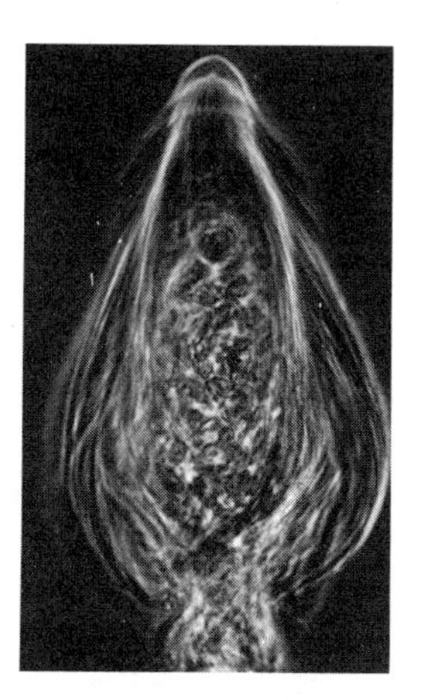

FIGURE 8-17. *Trichonympha gracilis,* 230×.

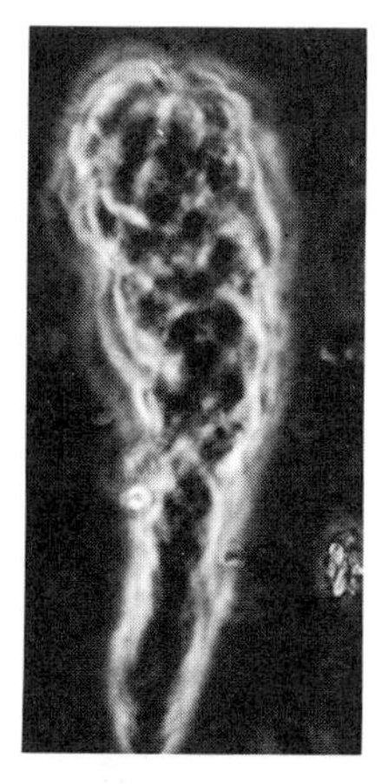

FIGURE 8-18. *Pyrsonympha vertens,* 230×.

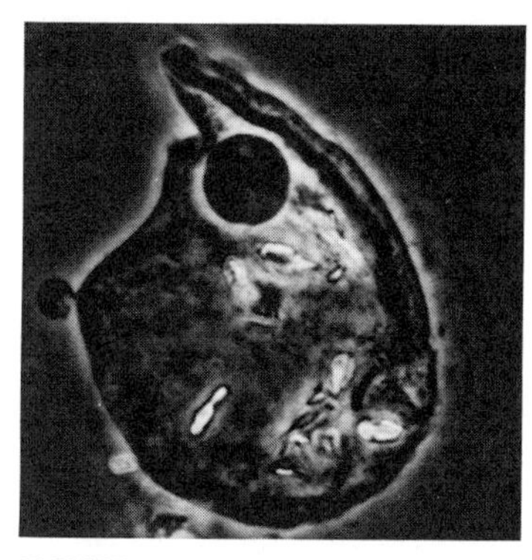

FIGURE 8-19. *Pyrsonympha major,* 340×.

FIGURE 8-20. *Cryptocercus punctulatus,* the roach's natural size.

If the existence of such weird creatures is a surprise, an even greater astonishment is in store for the naturalist who ventures into the hindgut of the woodroach, *Cryptocercus punctulatus* (see Figure 8-20).

Cryptocercus is a rare insect that can be found only in very few spots on our globe. In the United States, it occurs only in the forests of the Appalachian Mountains, in southern Oregon, and in northern California. Another species, *Cryptocercus relictus,* exists in Siberia.

Cryptocercus feeds exclusively on wood, but cannot digest it for the same reason as termites. In the woodroach, the digestion is also done by symbiotic flagellates in its hindguts. This dark brown wingless insect reaches a size of 30 mm when adult. Its colonies live in half-rotten logs of timber, digging tunnels as they feed. It cannot be found outside the log. Only when the trunk is consumed or does not provide sufficient protection does the family move out in search of another apartment. In its concealment, the insect has no enemies. Under these circumstances, it is strange that its distribution is so restricted. The reason is to be found in the climatic requirements of its protozoa. They have a very narrow thermal tolerance; they cannot survive hot summers or harsh winters.

The relationship between the roach and its protozoa is the same as that in woodfeeding termites. Neither the roach nor the symbionts can live without the other. As a safeguard against the loss of its protozoa, the insect has developed a complicated intestinal tract with a system of valves, which prevent their escape, be it through the anus or the hindgut entrance. The flagellates, on the other hand, had to solve the problem of making sure that the newborn nymphs are supplied with intestinal symbionts. The termite, as we have seen, stops feeding before the molting and the protozoa perish; but the nymphs are reinfected through proctolactic feeding. *Cryptocercus* found another solution: before the molting, the roach also stops feeding. Yet the protozoa don't die. Under the influence of the molting hormone that regulates the development of the insect, the symbionts too are affected. The protozoa, which reproduce asexually during the whole year, now undergo complicated sexual phases. Some encyst in solid cysts, others round up and just reinforce their cell membranes. These are preparations for the changed living conditions to be expected when the roach molts and the flagellates have to survive outside the host. The fecal pellets, which usually do not contain protozoa, now include cysts. When the nymphs hatch, they ingest the pellets. The cysts then hatch in the hindgut of the new host. The symbiosis is secured.

It was the late protozoologist, L. R. Cleveland, who did most of the research on these flagellates over a span of 40 years. His findings were unique. They disclosed entirely unknown life histories of the protozoa. It is not possible to elaborate upon Cleveland's findings within the scope of this chapter. The interested reader will find detailed descriptions in Cleveland's monumental volume on the woodroach,[3] as well as in many additional publications on this subject. At present, 26 different flagellate species belonging to 14 genera have been described and named (Yamin, 1974).[4] Actually, the intestines of this insect harbor many more forms. Cleveland rightfully concentrated his efforts on the large forms because their complicated and varied life histories were much more interesting to him than the small flagellates, which—to use Cleveland's own lively description—". . . fill the spaces between the apples and plums and grapes" of this incredibly crowded microcosm. Three of 26 known forms present in the roach's hindgut are shown in Figures 8-21a-c.

Actually, many more species do exist in this roach's hindgut. One of the most remarkable organisms that I found during my visits to Mountain Lake, where most of the research on this insect has been done in the past, is the one pictured in Figure 8-22a. I encountered it in one of the first roaches I dissected. From the beginning, it struck me as a most unusual animal. Yet it was not mentioned anywhere in the literature on *Cryptocercus*.

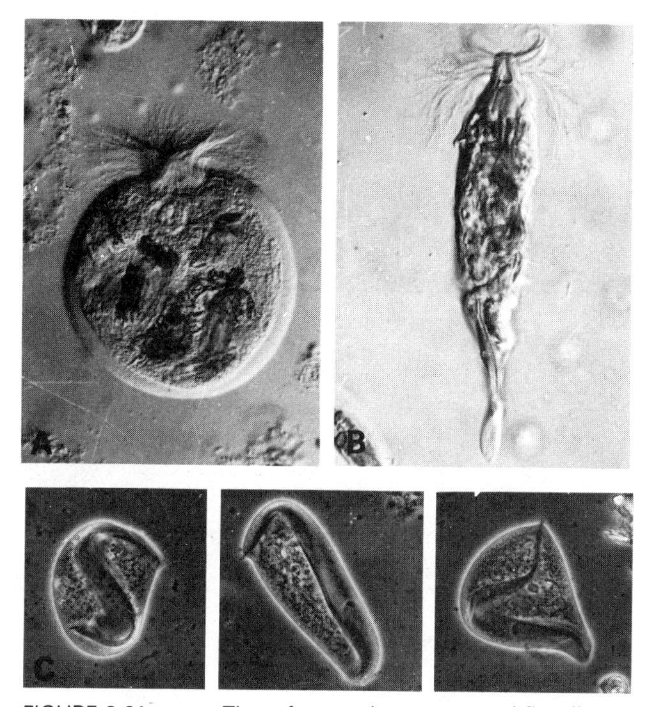

FIGURE 8-21 a to c. Three frequently encountered flagellates in this roach: (a) *Barbulanympha ufalula*, 190×; (b) *Urinympha talea* 275×; (c-e) *Saccinobaculus ambloaxostylus* 190×.

Further attempts to identify it proved futile. When I realized, after corresponding with Dr. Cleveland, that it was indeed unknown, it triggered future visits to Mountain Lake, in the hope of finding it again. This hope was not fulfilled. Instead, I discovered even more new forms, 31, to be specific, of which Figure 8-22 shows only a small selection.[5] For my observations, only about 100 roaches were available. If the search had been continued, additional species could have been detected.

Another interesting case is recorded in Figure 8-22b. This cell was vibrating at a speed that made it impossible to make out its shape, much less to focus it sharply. The only way to photograph it was to release the flash at random. The few exposures in the sequence shown give, however, some information on the general shape of the animal, the position of the nucleus (arrow), and the partly extruding axostyle. It is a small cell about 50 microns (50/1000 mm) long. Without the use of an electronic flash, it would still be "unseen." An attempt was made to fix the cell by adding a drop of 2% Formalin under the coverglass, but either it was lost or it rounded up on fixation and could not be recognized.

Figure 8-22c is another of those forms that have been detected only once. While most of the organisms inhabiting the roach's intestines are extremely lively, this one is rather phlegmatic, changing its body shape from

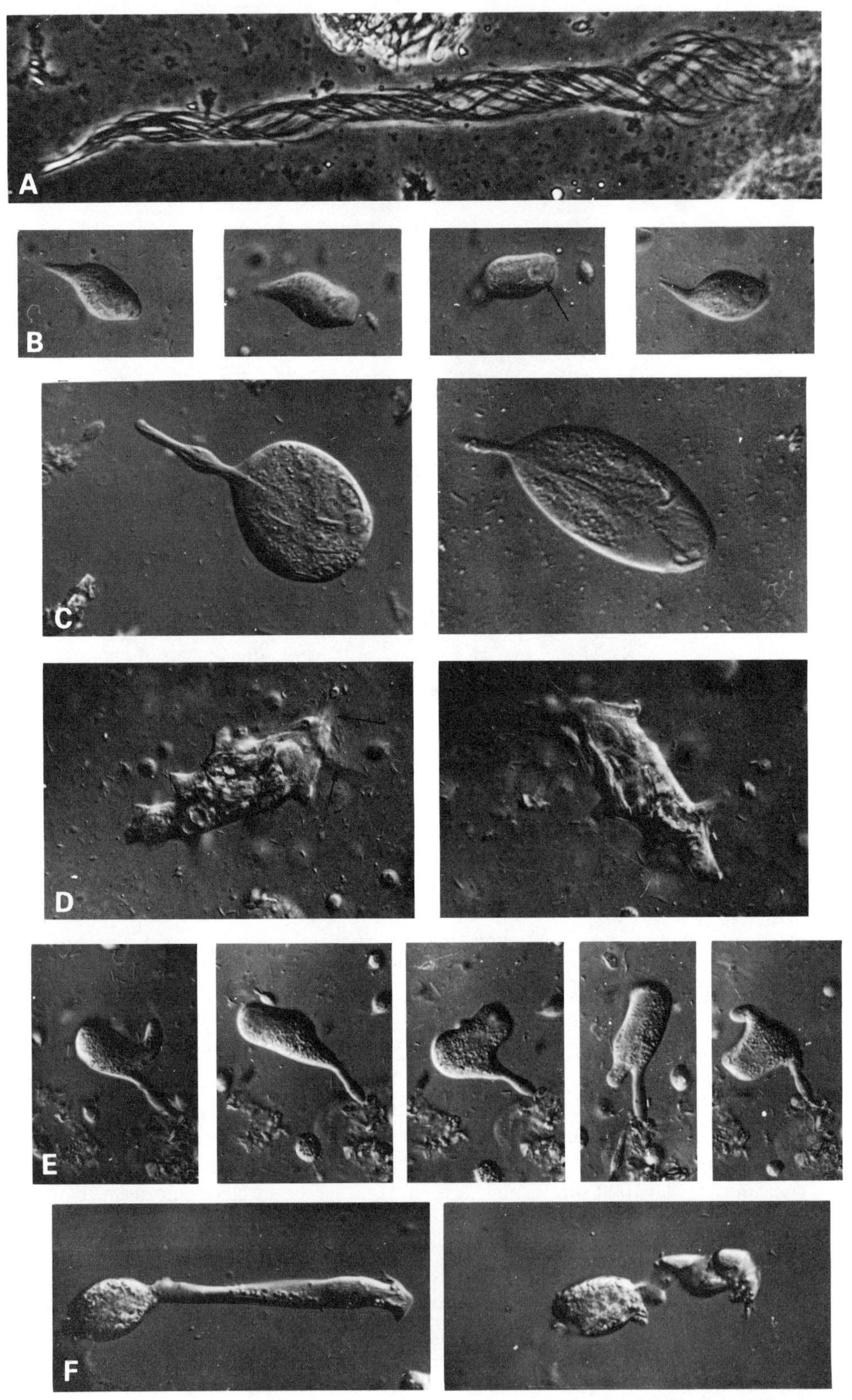

FIGURE 8-22, a to f. Selected, so far unreported, species in the roach's hindgut, 260×

round to oval and back again, with its axostyle protruding more or less accordingly.

Figure 8-22d is one of the oddest creatures I have ever seen in the microworld. It is hardly possible to make out head and tail. As a matter of fact, it has two bundles of flagella at the front part of the cell (arrow) that it uses to drag itself, clumsily and slowly, over the slide. The flagella are not well seen because the focus in on the peculiar bulges of the body surface. The flagella are used as a swimmer uses first one arm, then the other for propulsion.

The majority of as yet unknown organisms in *Cryptocercus* are characterized by a peculiar organ, the axostyle. This serves partly as a support of the cell. It is also partly instrumental in activating the erratic movements these animals exhibit, tirelessly contracting, expanding, bending, turning left, turning right, never resting. Figures 8-21, of three *Saccinobaculi,* show such a prominent axostyle and the way it acts. Another animal (see Figure 8-22e) uses the tip of its axostyle to attach itself to a detritus flake, settling down and arbitrarily changing its body shape. But in Figure 8-22f, another cell only stretches and expands in an abrupt, but monotonous way, with an absolutely grotesque appearance at the contracted stage.

Are these small forms beneficial to the roach? Are they symbionts, parasites, or commensals? A striking observation is that they do not (with some exceptions not treated in this chapter) contain wood particles. This can be easily established by looking at them with polarized light. Since wood polarizes strongly, it stands out clearly at extinction, if present. If these flagellates do not feed on wood, they must get their nourishment by absorbing fluid through the body surface. In other words, they cannot be considered to be symbionts, but are probably commensals because they apparently do not harm the host.

But then the function of the axostyle can be explained in a different way. When the animals are observed, they tend to stay in place and twist and writhe. If the axostyle in these animals has the function of locomotion, which is generally assumed, it can be only of secondary importance, inasmuch as there is not much space to get around in this unbelievably crowded environment. The primary function of this organ could be to flex the cell body constantly, thereby facilitating the absorption of the nourishing intestinal fluid in which they bathe. If the commensals were motionless, the absorptive process on which they depend, called osmosis, would be impaired in the cramped environment of the hindgut. The action of the axostyle would therefore free the individual flagellate from its competing neighbors, exposing its body to absorption.

Considering the rarity of the woodroach, few naturalists will have the opportunity to look at its intestines. Yet termites are abundant everywhere in the tropical, subtropical, and moderate climatic belts of our globe. To get at the flagellates is not too difficult. The tools needed are a scalpel; two tweezers, one with a sharp point (#5); a watch glass; a magnifier, to be held in the eye as watchmakers use them; and a 0.4 to 0.6% solution of sodium chloride. First, decapitate the termite with the scalpel. Then, using the dull tweezers to hold the insect at its thorax, the part of the body between head and abdomen, insert the sharp tweezer into the anus to pull out the intestines, which are immediately transferred to the previously prepared salt solution in the watch glass. By shaking the intestines in the fluid, the flagellates will flow out. The smaller the salt drop, the denser the flagellate population will be. One or two pipette drops of it are then transferred to the slide for observation.

Protozoa in the Rumen of Ruminants

Another hiding place where protozoa have made their home is in the rumen of cattle, goats, sheep, and other ruminating animals. The rumen contains, aside from bacteria, huge numbers of strange ciliates.

Ruminants have digestive problems. They must take in great quantities of food that is poor in terms of energy and difficult to digest. Plant cells have a cell wall of cellulose. As long as this wall is not broken down, the digestive process cannot convert the nutritious content of the plant. But ruminants lack an enzyme capable of breaking down cellulose. A few decades ago it was assumed that the ciliates in the rumen were capable of digesting cellulose and could even transform it into albumin, which would then be useful to the cow or goat.

Today we know that this is not so. It is not the protozoa that break down the cellulose, but the cellulose-fermenting bacteria present in enormous quantities in the rumen and reticulum. The contents of these compartments of the ruminant's stomach undergo fermentation. Without these bacteria, which are true symbionts, ruminants indeed could not live. The protozoa do not and cannot do the job; on the other hand, they are in no way harmful. Indeed, they are even useful, to a certain extent, as we shall see.

The lifespan of ruminant protozoa is extremely short—one or two days only. When the ingesta pass into the third and fourth compartments of the ruminant's stomach, and finally into the intestines, the ciliates are digested. Since half of the ingesta pass through the first two compartments of a cow's stomach daily, a single individual may live only for a few hours.

One would think that only the simplest cell structure would do for such a short lifespan. Yet these animalcules are, among the one-celled organisms, some of the most developed and complex in the microworld. They have a rather rigid cell wall, and some species have skeletal plates on their ventral side, structures that serve to reinforce the cell body (see arrow, Figure 8-23a).

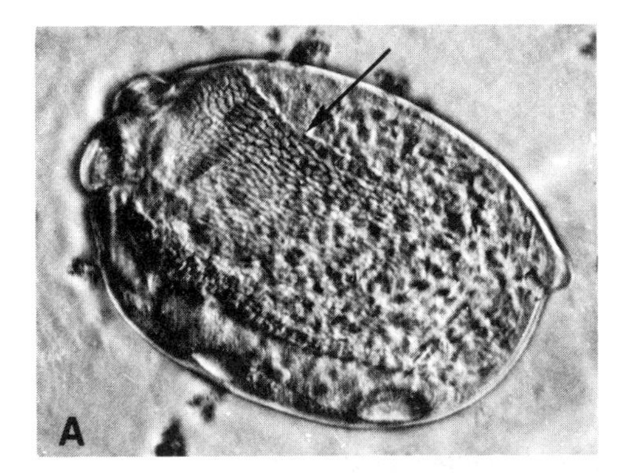

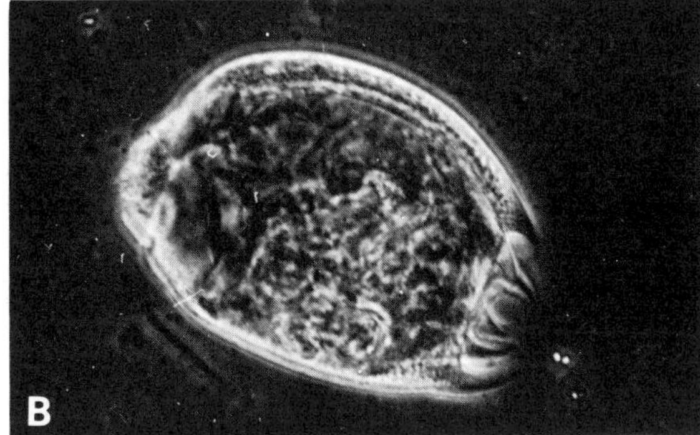

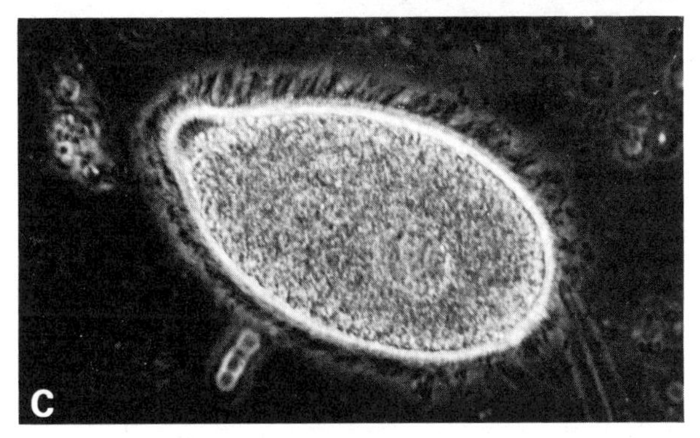

FIGURE 8-23, a to c. Protozoa in the rumen of ruminants: (a) *Diplodinium bursa,* 480×; (b) *Diplodinium dentatum,* the species name of "dentatum" (toothed) refers to the thorny processes, 550×; (c) *Isotricha prostoma,* 620×.

There are two families of ruminant protozoa, the *Ophryoscolecidae* and the *Isotrichidae*. The *Ophryoscolecidae* have no cilia except at the anterior end, where there are one or two ciliated zones that serve for locomotion and feeding. Some species are excellent swimmers and may achieve a speed of 20 times the length of their cell bodies per second.

The *Ophryoscolecidae* are the only family among the ciliates that has developed a kind of digestive tract. The mouth opening, or cytostome, leads into a sac formed by a thin membrane, where the food (bacteria, plant particles) is collected. At the back of this sac is an opening, called the cytoproct, through which undigested food particles are eliminated. Peculiar organs are the toothlike structures some species have at their posterior end; their function is unknown (see Figure 8-23b).

Like other ciliates, the *Ophryoscolecidae* reproduce vegetatively through fission. But their sexual reproduction or conjugation shows some interesting peculiarities. Since the conjugating pairs are connected at their mouth openings, the wandering nuclei must migrate quite a distance in order to fuse with the stationary nuclei inside the conjugating partners. For this reason, the wandering nuclei have assumed the shape of spermatozoa: They have a head and a flagellum for locomotion—a unique development in one-celled animals.

These "primitive" creatures have evolved a system of myonemes (contractile fibrils), which serve to contract and release their cilia and to control the opening and closing of the mouth or anus. The myonemes, for their part, are connected with a control center, a kind of simple nervous system. All told, these ciliates are an unbelievably complicated creation combined in one single cell for a minimal lifespan.

The preservation of life is not only a question of reproduction; it is also necessary to ensure favorable living conditions for the offspring. How do the ruminant protozoa get from one host into the other? A newborn calf has no ciliates and remains free of them as long as it is on a milk diet. Only when the calf switches to a plant diet will it be infected with protozoa. This is achieved while the herd is feeding in a meadow. The saliva of the adult animals contains ciliates from regurgitated ingesta. Some of this saliva becomes attached to the grass blades, and when a calf grazes over an area previously visited by an adult, it takes up some of the protozoa. It is being "inoculated." The ciliates then quickly reproduce in the calf's rumen, provided, of course, that the animalcules were able to survive in the hostile environment outside their host.

In hungry, weakened cattle, the number of ciliates declines rapidly. It is possible to eliminate the protozoa population by letting the animal fast for six or seven days without water. (Reduced feeding with water and the addition of one liter of milk a day is also effective.) These experiments have shown that the ruminant protozoa are in no way essential to the digestive process of the cow. A termite, deprived of its flagellates, must finally perish because it is unable to digest the wood without its symbionts. Ruminants, on the other hand, are no worse off without their ciliates.

Very probably, the protozoa, being ultimately digested, provide their host with a source of protein. This is the price for a hospitality of one or two days, during which the ciliates helped themselves to some of the starch that the ruminant's food contained.

Figures 8-23a and b show representatives of the family *Ophryoscolecidae*; Figure 8-23c shows a member of the family *Isotrichidae*.

Pilobolus, the Sharpshooting Fungus

When next you go for a walk in a meadow where horses are grazing, don't turn aside from the horse dung balls at your feet (see Figures 8-24a and b). On them may be found a tiny fungus named *Pilobolus crystalinus,* which is well worth investigating.

FIGURE 8-24a. A dung ball of a horse on which hundreds of *Pilobolus* fungi are growing. Reduced to about ⅔ of natural size.

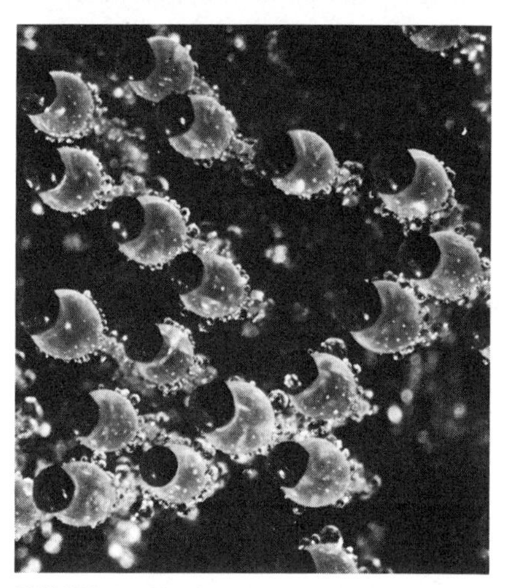

FIGURE 8-24b. Close-up view of the same dung ball slightly magnified, 11×.

FIGURE 8-24c. *Pilobolus crystallinus* as seen through the microscope, about to shoot off its cap. The function of many droplets is not known, 35×.

Pilobolus consists of a very tender, slim stem, which maintains an oval body, the subsporangium. This stem is topped by a small black cap, the sporangium, or spore case. Where stem and sporangium join, an area of reddish orange pigment is visible. The entire fungus is covered with countless tiny droplets (see Figure 8-24c).

With these simple means, *Pilobolus* performs a unique spectacle of power, skill, and precision. While observing it inside a jar with a magnifying glass, one may hear a very weak sound, as if small objects were hitting the glass. And, indeed, minute black dots appear on the inside of the jar as *Pilobolus* starts firing its spore cases—the containers of future *Pilobolus* generations.

In the meadow, these spore capsules are supposed to land on grass blades, which later on may be ingested by plant-feeding animals. *Pilobolus* relies on the cooperation of the horse, the cow, or any other herviborous mammal for its distribution. The animals swallow the spore cases, which withstand digestion; indeed, their germination is stimulated and accelerated in the warm surroundings of the host's intestines. Then, a few miles later, the spore capsules are eliminated again to resume the life cycle. But nature is not satisfied to let the spores drop on the manure where *Pilobolus* grows—they must be shot off, as far as possible. To achieve this, the spores shoot only in the direction of highest light intensity, because where there is light, there is open space and little likelihood of encountering obstructions.

How *Pilobolus* finds the direction of highest light intensity was long a puzzle, until a Canadian mycologist, Reginald Buller, proved that the swelling body or subsporangium that carries the spore case fulfills the function of a lens or simple "eye." The swelling body is a cell filled with clear cell sap. It is built in such a way that the light rays that enter it converge at the funnel-shaped base, as do the rays that enter our eye and meet at the retina. The base of the swelling body, with the reddish-orange pigmentation, is a part of the plant that is especially sensitive to differences in light intensity. Suppose sunlight strikes the little fungus from the side. The entering light rays are deflected and concentrated in such a way that one side of the "funnel" receives more light than the opposite side. As a result, the more intensely illuminated area receives an additional stimulus that affects and accelerates growth on that side. *Pilobolus* turns until it points to the light source, because in this position the sensitive area at the base of the subsporangium is uniformly illuminated from all sides.

Pilobolus has an excellent aim. But this isn't all. It also considers the missile's fate after it has been fired. What would happen if the spore case bounced off the grass blade it hit? It would drop to the ground, and the chance of an animal ingesting it would be small. To prevent this, a minute droplet of sticky cell sap accompanies each spore capsule on its flight and glues it to the grass blade it strikes. The fact that the cap is rounded also has a purpose. Should the spore capsule strike a blade on its rounded side, which is not sticky, it could bounce off and be lost. The curved shape of the cap makes it possible for the missile to turn around on impact and contact the target with its flat, sticky side. Besides, this glue is of such exceptional quality that even rainwater cannot wash it off.

If you want to watch *Pilobolus*, whether in a meadow or in a jar at home, you have to get up early in the morning. *Pilobolus* emerges from the dung during the night and is almost fully developed by the time the sun rises. It seems to be in a hurry to be ready for its barrage around 10:00 A.M. Why? At this time of day, the sun is at an angle of approximately 45 degrees over the horizon. If *Pilobolus* fires its cap at this angle, it has the chance of reaching the greatest possible distance. Millions of years before humans thought of shooting an arrow, *Pilobolus* "knew" the basic laws of ballistics.

Where does the energy originate that is required to shoot off the cap? It's the swelling body that fulfills this important function. Again, the optical properties of the sporangium pull the trigger. If the plant is oriented toward the light source, which is the sun under natural conditions, some of the rays are cut off by the black cap. Those rays that do enter the cell are condensed inside the sporangium. This results in photochemical reactions that cause the cell to swell until it bursts at its weakest point, the top. The spore case is then squirted out.

According to Reginald Buller, who has worked out *Pilobolus*'s life history, the pressure inside the sporangium immediately before the explosion amounts to about 5 atmospheres. Vertically, the cap can reach a height of two yards, horizontally about 2½ yards—a remarkable performance for a plant which, when fully grown, reaches a height of only 3/8 of an inch. After the cap flies off, *Pilobolus* breaks down, its reproductive function accomplished.

In its relation to herbivorous animals, *Pilobolus* is a typical commensal. It takes full advantage of its host without harming it, but also without giving anything in return. It is fully dependent on the horse, the cow, the elephant. They provide the substrate on which it grows; they provide the environment in which its spores can germinate; and they provide the means of transportation for its offspring which, over millions of years, have assured its cosmopolitan distribution. It certainly belongs among the most marvelous plants, alongside the Venus's flytrap and the bladderwort.

Pilobolus is easily grown at home. After you have obtained fresh horse, elephant, or cow dung, sprinkle some water on it with a pipette—a procedure that should be repeated every evening or every other evening. A little water should always remain at the bottom of the jar to keep the air inside humid. After covering the container with a glass plate, put it on a window ledge, but not in direct sunlight. To secure some ventilation, place a rubber band around the glass plate; this is sufficient to prevent the plate from closing the jar too tightly.

If you look at the culture early in the morning after three to five days, you will notice a great number of glittering silver whiskers, all pointing the same way, toward the light. This is *Pilobolus*, which has begun its lifespan of a few brief hours.

Fascinating experiments have been done with *Pilobolus*. Its sharpshooting ability can be tested by growing it in darkness under a cardboard box. If an opening is cut into the top of the box or on one of the side walls, then covered with a piece of cellophane, most of the spore cases will land on the cellophane. The windows can also be provided on opposite sides of the box: *Pilobolus* will select one of the two windows as a target, but will not shoot into the dark space in between. Its accuracy will depend on the intensity of the light that enters the box. The light can be regulated by placing electric bulbs at different distances from the openings in the box.

That the swelling body is in fact a lens capable of producing an image can be shown in a simple way by anybody with a microscope at his or her disposal. The instrument is placed in front of a window at a distance of two or three yards, with the condenser removed. Select a single fungus from the

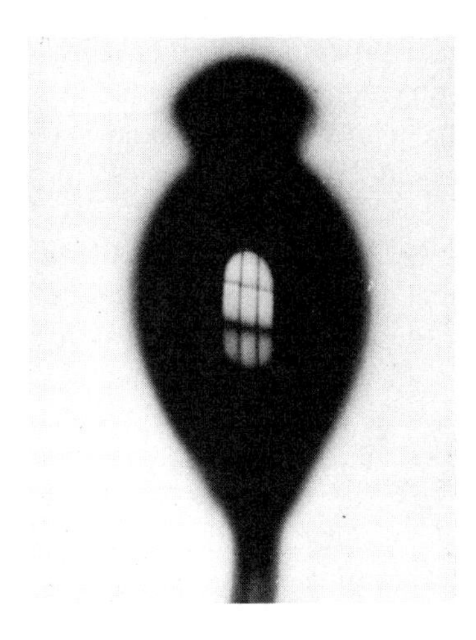

FIGURE 8-25. The sporangium is actually a lens capable of forming an image of objects before it. For this photomicrograph, the microscope was put in front of a window, 90×. (Based on an idea by Reginald Buller, who proved with a photomicrograph of the letter A that the sporangium is indeed a lens.)

culture and remove the droplets by cautiously going over the subsporangium with a soft hairbrush. With a tweezer, transfer the piece of dung on which the selected fungus grows to a slide, and arrange it so that the plant is horizontally positioned. The swelling body must not touch the slide. Adjust the flat side of the mirror so that the window appears as a distinct image (see Figure 8-25). This experiment requires speed, and some practice, because the fungus, especially in warm weather, fades within a few minutes, even if transferred on a piece of dung. A 3.5× objective and a 10× ocular are best for this observation.

Some Hitchhiking Plants and Animals

It was a cruel blow for the gourmets of the seventeenth century when the "curious philosophers" of the time examined the appetizing cheeses on their dinner tables through the microscope. There they found a strange guest, of which Pierre Borel wrote in his "Observationum":

> *The Acarus reminds one of a hairy bear or rather a terrifying Porcupine. It is really marvelous that the atom . . . which the animal occupies can have so many organs. One can see in them feet, eyes, nerves, in short all parts of a living being. Admire, therefore the power and wisdom of Nature, dear reader. She has with much skill created many things into a very small space. She has, without difficulties, provided small animals with as many organs as she did whales or elephants.*[6]

Borel writes about the cheese mite, now known scientifically as *Tyroglyphus siro* L. (see Figure 8-26a). Its offspring emerge from eggs laid by the fertil-

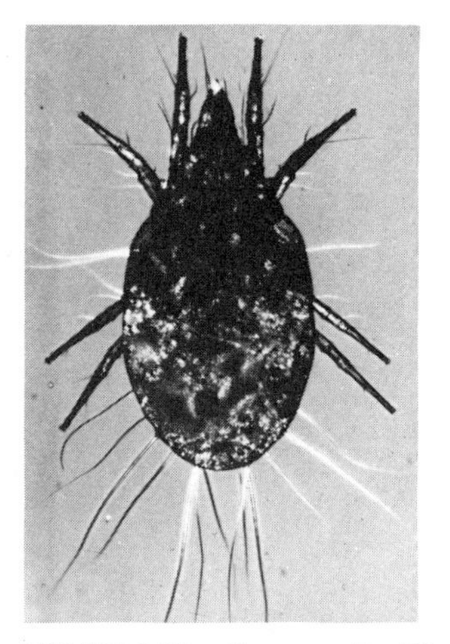

FIGURE 8-26a. Cheese mite, 73×.

FIGURE 8-26b. *Hypopus* of cheese mite, 80×.

ized female and are distributed at random while she is crawling about untiringly. When the larva hatches, it already shows the characteristics of the mature mite (the imago), with the difference being that it has only six legs, while the adult animal boasts eight. The six-legged larva molts and develops two more legs, this way reaching the first nymph state (protonymph), It skips the second nymph state and goes immediately through the third, becoming the adult mite. This is the normal way things go in the cheese mite's world. But sometimes the protonymph does not choose the direct route to the adult form. Instead, it assumes a completely different appearance, changing into a creature called *Hypopus* (see Figure 8-26b) which, with its short, plump legs, can hardly move and, because it has no mouth, cannot eat. But at the tip of its abdomen, it has several circular structures. These are special organs, sucking discs, which allow the nymph to attach itself to insects or other small animals that happen to pass by. In this manner, the nymph gets a free ride to some other location. When *Hypopus* "feels" that such a new and promising environment is at hand, it jumps off its vehicle, changes into the third nymph state, and then completes its metamorphosis.

The *Hypopus* stage has also been called the "wander-nymph" or the "traveling dress" of the *Tyroglyphidae*—and it has caused zoologists considerable headaches. Because of its completely different appearance, it has been considered to be a separate genus of mites, or the male form of the *Tyroglyphidae,* or even a parasite. Only at the end of the nineteenth century was the life history of this remarkable animal finally unraveled. Overpopulation triggers the protonymph to change into a *Hypopus*, but the mechanism that induces this detour in the development of the cheese mite is still a mystery.

Today the cheese mite has been driven from our cheese plates, but exists successfully in nature—whether in the woods or in garbage collec-

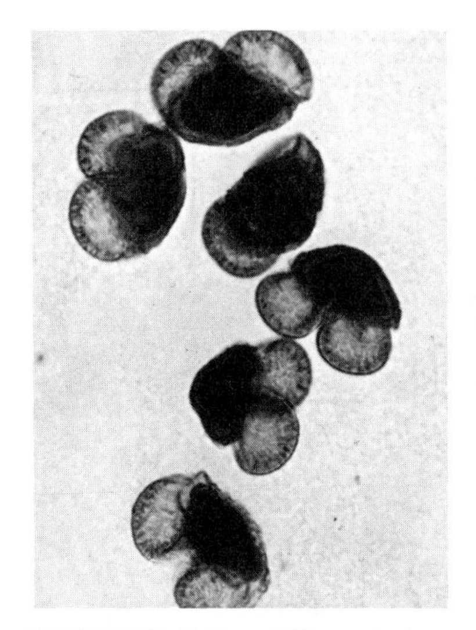

FIGURE 8-27. Pollen of *Pinus strobus,* 300×.

FIGURE 8-28. Seed of dandelion, 4×.

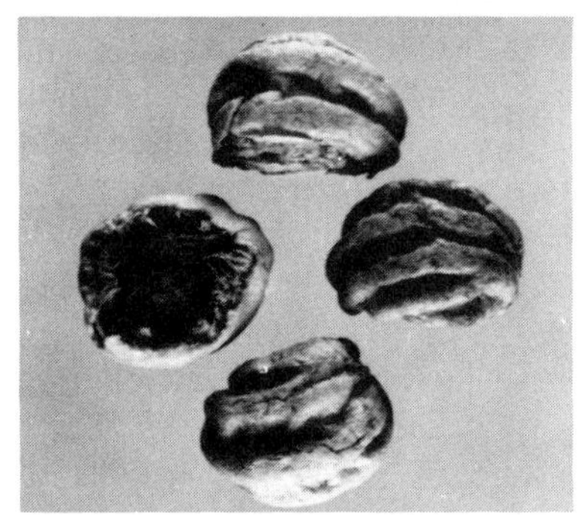

FIGURE 8-29. Seed of Indian cress, 3×.

tions. Despite its common name, its diet is not restricted to cheese. It enjoys a healthy appetite and feeds on all kinds of decaying or dry substances of animal or plant origin. It appreciates flour, sugar, ham, dried fruits—even mattresses and other objects of our daily life, which is why we still consider it an enemy.

But wind and water are the more common hitchhiking vehicles in nature. The pollen grains of pine trees must function high up in the treetops; they are therefore provided with two air sacs, wings that may carry them for many miles (see Figure 8-27). The dandelion seed is attached to a parachute that can never fail because it is already opened (see Figure 8-28). The seeds of the Indian cress, on the other hand, build a life preserver around their kernels—a layer of corky material that makes them drift on the water (see Figure 8-29). This plant originally came from Peru, where it still grows near streams and rivers.

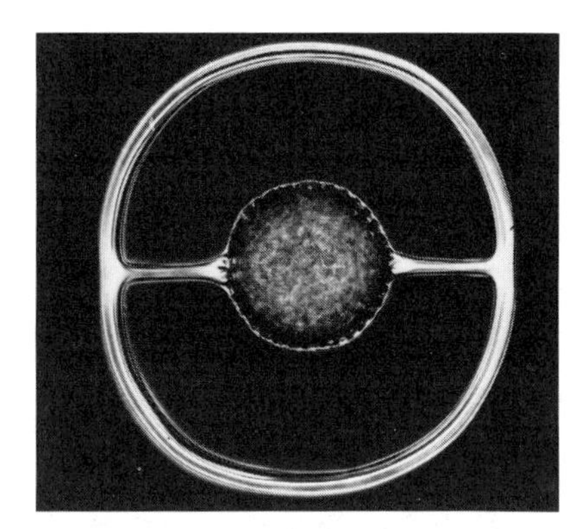

FIGURE 8-30. *Saturnulus elipticus,* a radiolarian, 190×.

FIGURE 8-31. Seed of *Celandine* with elaiosomes, 23×.

An even more distinct life preserver has been developed by a minute oceangoing animal of the Radiolaria, *Saturnulus elipticus.* (See Figure 8–30.) The delicate ring around its body keeps the animal floating near the surface of the sea, where it finds its food among other microscopic organisms.

These Radiolaria are some of the most beautiful creatures of the microworld. They consist of siliceous skeletons that are indestructible unless damaged by mechanical force. After the animal dies, the living matter decays, but the skeleton remains and gradually sinks to the bottom of the ocean. The animal pictured in Figure 8-30 was alive a long, long time ago. It was found in the ooze brought up from a depth of two and a half miles by a scientific expedition to the Indian Ocean.

Some plants attach their fruits to the fur of passing animals. The burdock's prickly, barbed burrs have a way of getting caught in stockings. The hooked burrs serve a double purpose: They prevent animals from eating the fruit, and they render the plant a first-class hitchhiker.

For the oak tree, a squirrel is the vehicle. It helps itself to some of the dropped acorns for food and carries away and hides others, which it then forgets about.

An even more refined way of hitchhiking evolved in the tiny seeds of the celandine. These seeds have an appendage, known as an elaiosome, that has nothing to do with the germination of the seed (see Figure 8-31). It consists of an albuminous substance, which ants love for food. They collect the seeds, carry them into their nests, feed on the elaiosome without harming the seed itself, and discard it again outside the nest. A truly shrewd way of getting a ride!

9

Some Aspects of Behavior Revealed by the Microscope

When Stentor Gets Scared

Human behavior is often extravagant. This is nothing new, and the effort to explain our strange reactions has been a subject of uninterrupted studies ever since humanity became conscious of itself. Yet, peering through the microscope, observers have discovered reactions among the lower animals that can be described at least as inconsistent. Although we by no means act logically at all times, we usually expect the lower animals to behave within a fixed pattern outlined for them by nature.

Any microscopist who has ever observed the fauna of our puddles, ponds, or vases knows how sensitive these animals are. If a rotifer (see Figure 9-1), also known as the wheelanimalcule, is disturbed during its meal—should, for example, an infusorian without any ill intentions collide with it—the rotifer immediately contracts. The consciousness of pending danger is certainly an ever-present instinct in the inhabitants of the microworld. Yet their memories are very short. As soon as the danger passes, the wheel-animalcule will stretch out again and resume playing with its cilia—until the next alarm.

Many more examples of such shock reactions could be given. The irritation can be of different kinds. A strong ray of light stops the movement of an amoeba's pseudopod. A mild electric shock affects almost any microorganism. Yet living things do not always follow rigid laws. The behavior of one-celled organisms is often much more complex than is apparent. An interesting example of this complexity is *Stentor coeruleus*, the trompetani-

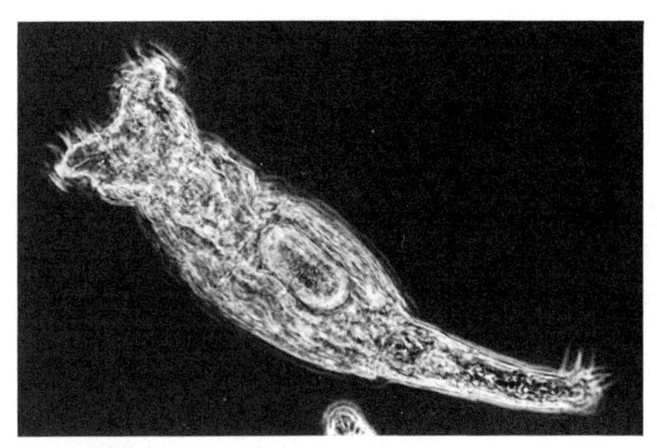

FIGURE 9-1. A *Rotifer.* The oval structure in the center is the stomach, 125×.

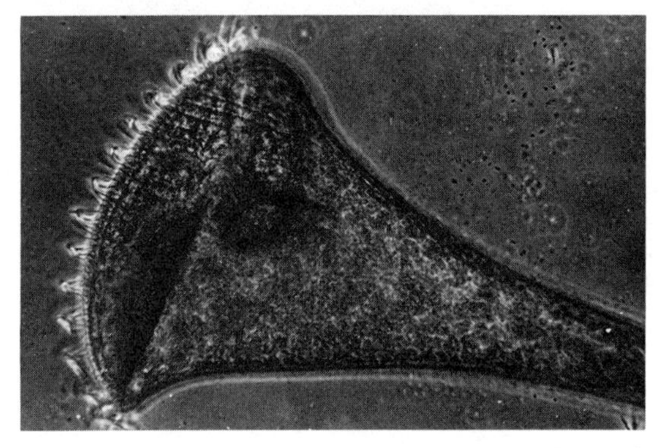

FIGURE 9-2. *Stentor coeruleus,* the trompetanimalcule, 185×.

malcule. Most of the time we find it expanded, displaying its elegant figure (see Figure 9-2), exercising its cilia in order to cause a vortex to flush bacteria, small algae, and infusoria into its cell body. If we disturb the animal by lightly tapping the microscope stage, it contracts immediately but, like the rotifer, will expand again shortly. If we repeat the disturbance, it is possible that the animal "gets scared" again and contracts, but it may also be that it does not respond to the irritation. The reaction may cease at the second, third, or fourth disturbance or even later. The point is that *Stentor* does not react automatically (see Figures 9-3a, b, c, and d).

The type of irritation selected in this case is, of course, not the best. A vibration caused by tapping is something *Stentor* never experiences in nature. This type of vibration was chosen here only because the attempt to photograph the shock reaction required the use of a coverglass; the mischiefmaker therefore was unable to touch the animal directly.

A more natural method was tried at the turn of the century by the American zoologist, H. S. Jennings. Among other ways, he irritated the animals at their mouth openings with a fine glass needle, thus imitating the

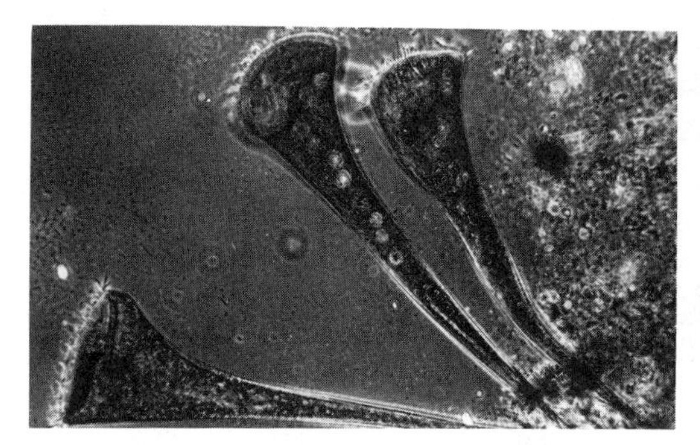

FIGURE 9-3a. Group of *Stentors* in feeding position attached to a detritus flake, 65×.

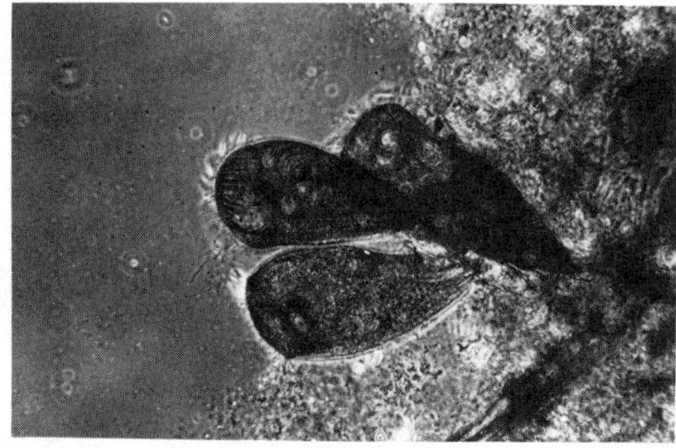

FIGURE 9-3b. When disturbed, the animals contract, 65×.

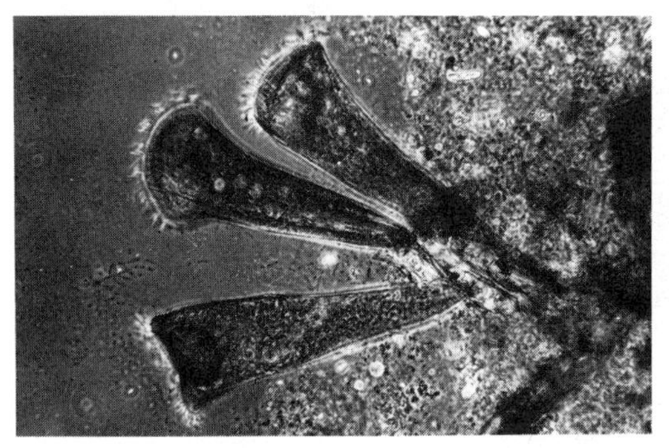

FIGURE 9-3c. Soon they expand again, 65×.

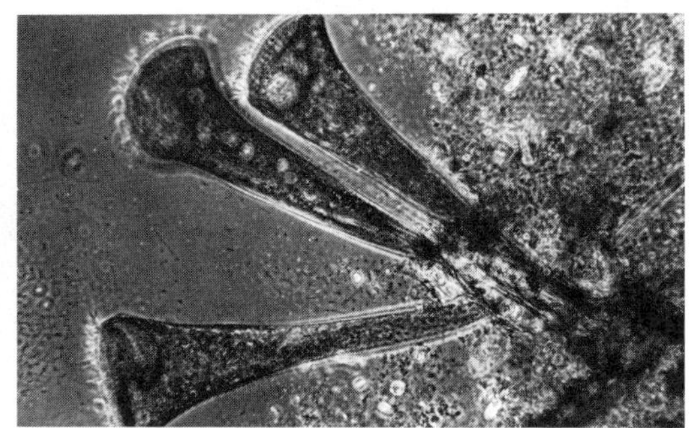

FIGURE 9-3d. The animals are now once more fully expanded and resume their play with the cilia, 65×.

collisions that so often occur in their natural surroundings. Jennings found that the animals reacted to the first half dozen touches by contracting. Then they ceased to contract. Several repeated touches were necessary to induce contraction; more and more "blows" with the needle were required. Finally, the animals "got mad" and took off.

It should be added that *Stentor* is not completely indifferent to repeated irritation before its patience gives out. Even if it does not contract, it tries to avoid the disturbance by turning away from the stimulus. A hungry *Stentor*, interrupted in its meal, will give up its moorings more quickly than a satisfied animal.

It is remarkable that *Stentor roeseli* is more difficult to chase than *Stentor coeruleus*. *Stentor roeseli* builds a little house for itself out of a gelatinous matter that its cell body excretes. It can retreat into this house in case of danger. *Stentor coeruleus* does not have this capability, but can attach itself temporarily to any object in the water and so escape if necessary.

What is going on in this ciliate that it behaves in such an inconsistent way? Why does it react differently to the same stimulus? Several possible explanations have been suggested.

The fact that the contraction ceases could indicate that the animal gets tired; but Jennings rejects this hypothesis as unlikely. It could be that the irritation is not perceived by the animal after some time, or that it is perceived in a reduced way, because the animal has become accustomed to it. But this explanation also has to be rejected. Although the animals do not contract, they do show a mild avoidance reaction, demonstrating that they do perceive the stimulus.

Jennings' own explanation is that the trompetanimalcule is capable of learning—it varies its behavior according to an experience. This experience tells the animal that the irritation that first made it react violently by contracting does not really disturb its activity, so it ceases to respond further. In other words, experience tells it that its violent response is useless because the disturbance does not stop.

If learning is understood as the modification of or adaptation to a mode of behavior, it must be concluded that *Stentor's* behavior is based on learning. Yet it is learning of very short duration because, after a period of undisturbed peace, the animal will react again with immediate retreat at the first of a series of touches.

Fish Scales

Who could fail to be intrigued by the exotic colors that some fishes, especially tropical ones, display? What is it that makes them so pretty? What gives them those intricate patterns and makes them glitter and shimmer at ever turn in the water? We can find the answer if we take a closer look at their scales. However, it would be cruel to pluck a scale from a live fish. Here the fishmonger at the corner comes in handy. He or she lets thousands of these little discs go down the drain every hour, never suspecting that they have a story to tell.

Fish scales are rewarding objects for microscope study, and they are easy to handle because no fixing or staining is required. Just put a scale flat on a slide with a little water, cover it with a coverglass or another slide to prevent it from curling up, and observe it with transmitted light.

There is little sense, though, in collecting scales from the chopping block. It would be like reading a book from which important pages have been torn out. Scales are set in the folds of a fish's skin. They are partly covered by skin tissue and when forcibly removed, the skin is destroyed so that we miss half of the story. It is therefore always advisable to obtain a piece of fish that has not been worked on.

A flounder is a good specimen to start with. The flounder scale, under low magnification, has a multitude of black spots of different sizes. Some

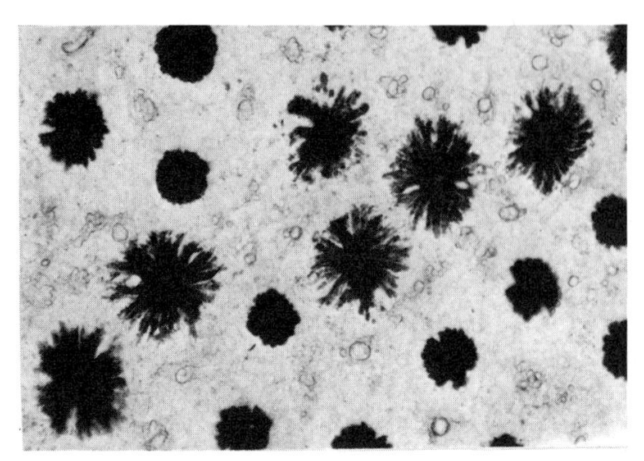

FIGURE 9-4. Melanophores from the skin of a flounder in different stages of expansion, 240 ×.

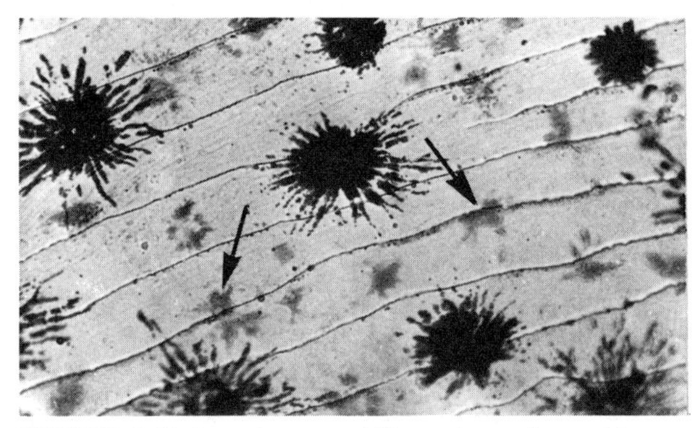

FIGURE 9-5. Melanophores and Xantophores (arrow) from the skin of a sardine, 240 ×.

are nearly round; others appear like inkblots, with numerous radiating processes. These are the so-called chromatophores, pigment-containing cells that are the basic structures of coloration in any fish.

If you happen to choose a gray flounder, you will find on its scales only black-pigmented cells, or melanophores, so named after the substance melanin, which forms the pigment granules (see Figure 9-4). But if you picked a brownish flounder, you would discover another type of chromatophores, known as xanthophores, after the substance xanthophyll. These xanthophores are visible as small, irregular spots of orange color. Other species of fish have yellow or red xanthophores, or a combination of these. The green color of the sardine, for instance, is achieved by a mixture of black and yellow chromatophores (see Figure 9-5). If the chromatophores of several colors are distributed over the skin—concentrated here, lacking there—an infinite number of patterns and shades is possible, as in some tropical fishes.

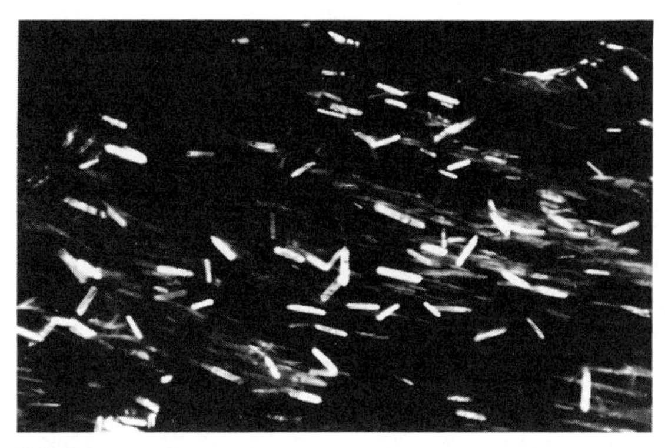

FIGURE 9-6. Guanin crystals from the scale of a salmon, 86×.

But the exquisite colors of fish are not only due to the presence of pigment. There is another substance, call guanin, distributed mainly beneath the scales, that is responsible for the irridescent effect so many fishes display. Guanin is in itself colorless matter, but it has the property of breaking up the light with a prismatic effect (see Figure 9-6). Goldfish in the aquarium are generously supplied with guanophores, as these light-reflecting structures are called. But we can find guanophores also by examining fish with a silvery appearance. Ask your fishmonger for a piece of salmon, and stress that you do not want the scales scraped off.

After cautiously removing a scale, put it on a slide, but this time upside down. Then add a drop of water, but not a coverglass, and adjust the lamp so that the light falls obliquely on the scale. If you now agitate the scale gently with the point of a needle, you will have a beautiful spectacle. To quote an enthusiastic microscopist, Philip Henry Gosse, who did this little experiment more than one hundred years ago and described it in his book, *Evenings at the Microscope:*[1]

> *The water around the mass is seen to be full of an infinite number of flat spicula or crystals, varying much in size, but very constant in form, a flat, oblong prism with angular ends. By reflected light they flash like plates of polished steel. But what appears most singular, is that each spiculum is perpetually vibrating and quivering; and independently of the rest, so as to convey the impression that each is animated with life. Every spiculum, as it assumes or leaves the reflecting angle, is momentarily brightening or waning, flashing out or retiring into darkness, producing a magic effect on the admiring observer.*

Most fishes have a given color pattern, which they keep for life. The surface feeders are all dark on the back and light on the belly; this gives them a certain protection against fish-eating birds that may attack them from above, as well as against predatory enemies from below. And because the catfish from the Nile has the peculiar habit of swimming upside down, it is its belly that is pigmented and gradually shades to a white back.

But many fishes have the ability to change their color, and some reach a surprising virtuosity in this respect. The flounder is one of these virtuosos. Put a live flounder in a tank with a chessboard as the bottom and it will try to imitate the pattern. It will not be able to produce geometric squares on its skin, but there will be very light and very dark patches all over its upper side. Black discs on white background or white discs on a black one are easier for it to reproduce.

How does the flounder do this? The answer is that its chromatophores are not stable. They have the remarkable ability to contract or to expand. If contracted, they give the skin a light shade; if expanded, a darker shade. Here the flounder has applied for millions of years a simple principle our newspaper printers thought out only recently. If, with low power under the microscope, you scrutinize the newspaper picture of a starlet whose new show has been announced, you will notice that its overall impression is achieved by numerous dots of different sizes (see Figure 9-7). The lighter regions of her pretty face are made up of tiny dots, the darker regions of heavy dots, which melt together. In the same way, the skin of a flounder shows chromatophores in different stages of expansion.

The stimulus for the flounder to change its color or pattern is received through its eyes. When the fish is blindfolded, by covering its eyes with tape, no color response results. If the flounder is kept with its head on a black background and its body on a white one, it will change to black. Here we have to stop for a moment and think of the intricate nerve system and the apparatus of tiny muscle fibers that can make possible the spreading or contraction of thousands of pigment cells according to what a fish "sees." The response is unconscious and automatic and even occurs when the fish is asleep, provided that its surroundings suddenly change.

The color change of the flounder has a definite purpose: It helps this fish to hide from its enemies. But there are other fishes in which the reasons

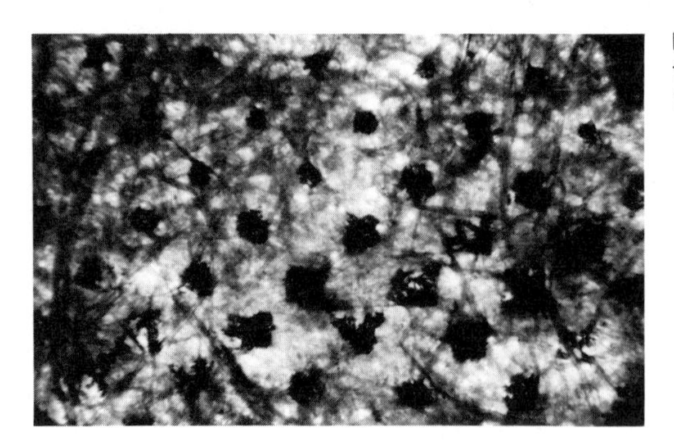

FIGURE 9-7. Halftone dots from a newspaper illustration, 21×.

for color changes are not so obvious. Some of the tropical fishes, especially those inhabiting coral reefs, have up to seven different liveries. They change their colors and patterns as they swim along, by no means always in accordance with their background. And, what is even more remarkable, they are able to effect the change within a few seconds. Are they doing it merely for pleasure? Experiments made in the New York Aquarium many years ago indicated that the color changes of these fishes are apparently made for emotional reasons and depend on their moods and activities. The colors changed when food was thrown into the tank, when the animals were frightened, when the electric light was turned on, when air was driven into the tank, or when other kinds of fishes were added to the tank.

Fish scales are interesting in still another respect. On its scales, a fish writes and carries around—in hundreds of copies—its personal life story for us to read, if we are able to decode its messages.

When a fish hatches, it has no scales. However, when it is one or two inches long—depending on the species—tiny plates appear, and their number at this early age remains constant throughout its life. The growing fish does not develop more scales, but the scales, once formed on its body grow with it to cover a greater surface. The individual scale increases in size through the addition of microscopic amounts of a bony substance to its edge. Just as a tree adds concentric rings to its trunk every year that can indicate its age when counted in cross section, so each fish scale adds rings that can be seen when viewed with transmitted light through the microscope. There is, however one difference. A fish adds several rings annually to its scales, and the number added depends on the life conditions it had to face.

In summertime, life is relatively easy. Food is abundant and the temperature is high, so a fish usually grows rapidly at this time of the year, and numerous, widely spaced rings are added to its scales. In wintertime, on the other hand, with temperatures dropping, the rhythm of life is considerably slowed, food is scarce, and the rate of growth is reduced. This fact is reflected by the addition of a few narrow rings to the scales. So, by counting summer and winter zones, we can estimate the approximate age of a scale and its owner.

A fish from the aquarium hasn't very much to tell us, of course. It leads a lazy life, is fed regularly, and remains at constant temperatures. The rings on its scales will be quite uniformly spaced and regular. But an adventurous fish, such as the salmon, has a great deal more to relate—like anybody else who does a lot of traveling.

To the experienced scale-reader, the scale from the Atlantic salmon, like the one shown in Figure 9-8, gives these particulars: that the fish spent two years in the river where it was born; that it then went out to sea to spend a year in the ocean; that it returned to the river in order to spawn; and that it went out to sea again—where alas—it was caught.

FIGURE 9-8. Scale of an Atlantic salmon, 10×.

The two river years can be recognized in the two zones of narrow rings in the center of the scale. During this period the fish grew slowly, owing to the scarcity of food in the river. The year in the sea is revealed by the sudden widening of the rings. Under greatly improved food conditions, this fish put on weight at a fast rate, and a larger body surface had to be covered by rapidly growing scales. Next to the sea-life zone, we notice a ragged line that goes all around the scale. This is another autobiographical inscription. It tells us that the fish has spawned, and it is called the spawning mark. In the process of spawning, the fish's girth is considerably reduced. Suddenly, the scales have to cover a much smaller body surface. Consequently, they become compressed and their edges become ragged and frayed. When the fish resumes its growth, a scar remains.

People may consider themselves fortunate that nature didn't provide them with scales that could reveal more about their intimate life than they would care to tell! It is not known how the mermaids and mermen feel about it.

10
A Trip to the Parks

The Hot Spring Algae of Yellowstone

A visitor to our national parks naturally pays attention to the objects of interest advertised in the brochures. At Yellowstone, the geysers and hot springs will absorb us, and with good reason: Nowhere else in the world can we find these phenomena in such numbers and variety.

However, for the microscopist, there is even more to be seen. He or she may wonder why so many of the hot springs and geyser drains glitter in all kinds of colors—green, blue, orange, brown, or yellow. This color symphony is caused by minute plants, the blue-green algae, of which we can find a great many different species if we know how to look for them.

Blue-green algae are primitive plants that are in their simplicity close to bacteria. They differ from the latter in that they contain chlorophyll, thus being capable of photosynthesis. Their cells are not as differentiated as the cells of higher plants. In the higher plants—the green algae, for instance, or other green plants—the chlorophyll is organized in a special structure, the chloroplast. In the blue-green algae, it is dispersed throughout the cell plasma. Blue-green algae also lack mitochondria, nuclear membranes, and chromosomes, though they do have DNA (deoxyribonucleic acid), the mysterious molecule that holds the genetic information for future generations. Blue-green algae reproduce only asexually by fission and by forming spores. They lack any means of locomotion, such as flagella or cilia, though some of the filamentous forms, like *Oscillatoria,* are able to move in a gliding fashion that we cannot yet explain.

Why is it that some blue-green algae thrive so well in hot waters? Each organism, whether plant or animal, has a specific heat tolerance. It dies when the plasma coagulates—a condition reached at different temperatures for different species. The tolerance for green algae or diatoms is about 50°C. Among the protozoa, the intestinal flagellates of the termites or the wood-feeding roach, *Cryptocercus punctulatus* (see Chapter 8), cannot survive a temperature of more than 37°C. Most amoebae die at 43°C. Yet the algae of the hot springs at Yellowstone live at temperatures of up to 77°C. Even higher is the tolerance of the "thermophilic" (heat-loving) bacteria, which can stand temperatures up to 88°C. If a microscope slide is left in a boiling spring at this temperature for eight to ten days and then examined, it is found to be densely covered with bacteria. (At Yellowstone water boils at 92°C because of the high altitude, 8000 feet.)

This high heat tolerance is not easily explained. One theory assumes that two kinds of protoplasm, the mycoplasm and the amoeboplasm, exist throughout nature. Mycoplasm is supposed to be highly resistant in general, needs little oxygen, and can stand temperatures of up to 95°C. Amoeboplasm, on the other hand, is thought of as more fragile. This theory postulates that higher plants and animals are endowed with amoeboplasm, while blue-green algae and bacteria are provided with mycoplasm. The weakness of this hypothesis is that there are many blue-green algae and bacteria that are made up in the same way as the thermophiles, yet have a much lower heat tolerance. The point must also be considered that the thermophiles perish if transferred to cooler waters.

Another theory points out that mitochondria are destroyed at temperatures of 48°C to 50°C. This is approximately the maximum that higher organisms can endure. Heat death may be explained by the destruction of the mitochondria, the essential organelles of cell respiration. Because blue-green algae do not have mitochondria, so the argument goes, they can survive high temperatures. However, there are many organisms that suffer heat death before the mitochondria can be affected. The deterioration of the mitochondria is not the cause, but rather the result of the cell's death.

How did the flora of the hot springs evolve? Some biologists believe that the ancestors of the thermophilic algae lived in waters of moderate temperature and adapted themselves gradually to the life in hot springs. Others argue that the thermophiles always existed in hot waters and that the "mesophiles"—plants that prefer cooler surroundings—descended from them. According to this view, the thermophiles are survivors from the very beginnings of life, when the average temperature of the atmosphere, the sea, and the inland waters was much higher than it is today.

Very recently, new information has become available that seems to favor the view that these bacteria are remnants of the time when life began on the earth, though adapting to become more moderate thermophiles. Two bacteriologists, John Baross from the Oregon State University and Jody Deming from Johns Hopkins University, reported a revolutionary discovery in the British magazine, *Nature* (Vol. 303, pp. 423–426) 1983, the existence of bacteria in volcanic vents deep on the Pacific Ocean floor, the Black Smoker bacteria. These bacteria live and reproduce at temperatures of 482°F (250°C), by far higher than in any of the hot springs in Yellowstone National Park. Until now it was believed (with a few exceptions) that the sun, through the process of photosynthesis, was the only source of energy that sustains life on earth. Yet these organisms subsist exclusively on inorganic matter, such as sulphur, manganese, and iron, in total darkness and under great pressure. The situation closely resembles conditions assumed to have existed in the primordial oceans at the time when life first developed.

In connection with these findings, a different interpretation is suggested by A. E. Walsby to explain the high level of temperature tolerance revealed by these thermophiles. Proteins are very complex compounds combining a number of amino acids, with a single protein molecule being made up of hundreds of amino acid molecules. Though the chemical composition of the Black Smoker bacteria shows that the normal amino acids are present, five amino acids "are of unknown identity." It is well possible that these unknowns will explain the extraordinary phenomenon of life at 250°C.

Yellowstone's blue-green algae occur either as one-celled organisms or as filamentous colonies. Figure 10-1 shows one of the unicellular varieties, *Synechococcus*. I collected it from an unnamed spring that attracted my attention because of its beautiful deep olive color. Its temperature was 70°C.

The most widespread are the filamentous forms. They can be found in almost all geyser outlets, where they often produce thick layers of cottonlike mats. If such a mat is teased apart with two needles, the microscope reveals a maze of entangled filaments (see Figure 10-2). This is the *Phormidium* alga.

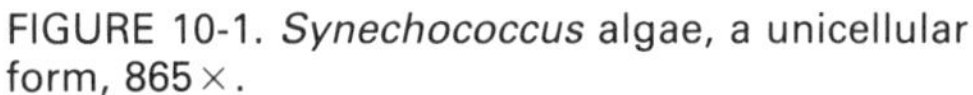

FIGURE 10-1. *Synechococcus* algae, a unicellular form, 865×.

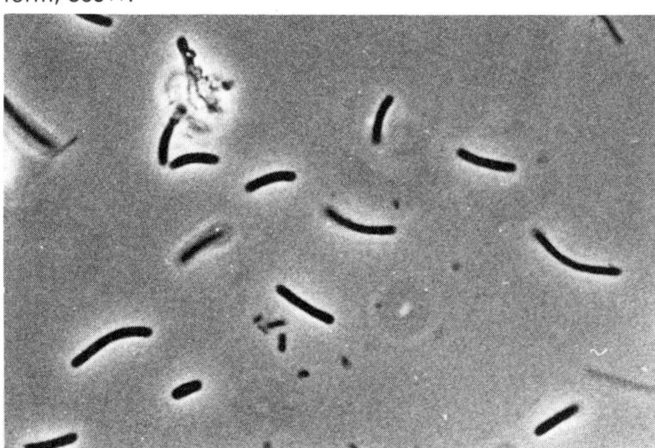

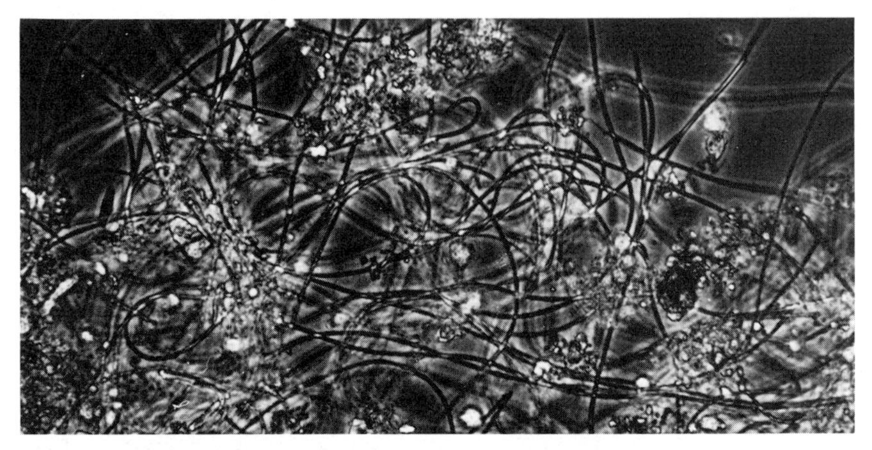

FIGURE 10-2. *Phormidium* algae, a filamentous form, 144×.

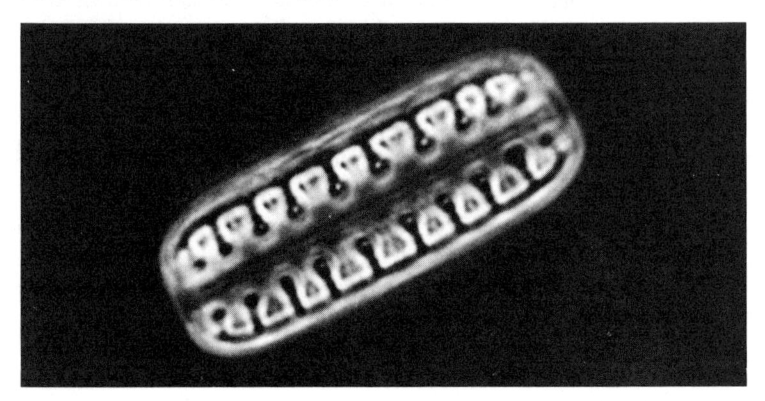

FIGURE 10-3. *Denticula thermalis,* one of the few diatoms that live in hot waters, 1200×.

If you wish to collect and preserve these algae on a trip to Yellowstone or at any other site where hot springs occur, you can fix them on the spot and study them later at home. They keep for a long time in 3 percent formaldehyde. Such a collection may also contain diatoms, some of which could have been carried into the water by air. Others, like the one shown in Figure 10-3 are among the few diatoms that have adapted to live in hot waters. It is certainly a beautiful piece of art.

The Petrified Forest of Arizona

An adventure of quite a different kind greets the traveler to Arizona's Petrified Forest, a unique area that contains many tons of colorful fossil wood lying on the ground, leading us back into the dimmest past of the earth, 160 to 170 million years ago.

FIGURE 10-4. Scene at the Petrified Forest of Arizona.

When Spanish explorers entered this area in the sixteenth century, they gave the name "Desierto Pintado," the Painted Desert , to the expanse right next to the Petrified Forest. Therefore, we can assume that they noticed the peculiar blocks strewn all over the place (see Figure 10-4), although they failed to mention them in their reports. Earlier, the region had been inhabited for centuries by Indian tribes. Traces of their civilization still exist in a number of ruins; the Agate House is one example, which also shows that these primitive aborigines knew how to use the agatized, or petrified wood for ornamental purposes.

In the nineteenth century, the American pioneers arrived to open up the wonders of Arizona to a broader public. Jim Bridger, the most original scout of the Wild West, told of strange things to be seen there. He had visited the petrified forest in Yellowstone National Park, where the fossil stumps are preserved standing upright, and reported having seen "petrified birds on petrified trees chirping petrified songs in the petrified air!"

When, in 1881, the Santa Fe Railroad began operations, the way was opened up to verify all of these strange stories. However, the way was also opened to crowds of vandals, who broke up the blocks with explosives in search of the quartz and amethyst crystals often to be found inside. Enterprising businesspeople soon discovered that a well-polished section of one of these crystals could get a good price. At the turn of the century, a manufacturer of sandpaper and abrasives even planned to establish a factory in order to crush and pulverize the material on the spot. This alarmed the conservationists, who submitted a petition to Congress to protect this unique area from further destruction. In 1906, during Theodore Roosevelt's presidency, the Petrified Forest was declared a National Monument, "for the benefit and enjoyment of the people." Although the damage done by former generations was considerable, some thousand blocks still survived.

The visitor naturally asks: What has happened here? When were these trees alive? How can they be identified botanically? What conditions made their preservation possible?

About 160 to 170 million years ago, toward the end of the era geologists call the Triassic period, today's Arizona was an extensive, flat valley into which numerous rivers discharged, forming huge lakes and swamps. These rivers carried along large quantities of mud, sand, and gravel, which piled up in thick sediments. Since the trunks of the trees in the "forest" have no branches or roots, and many even lack bark, it must be assumed that they couldn't have grown where we find them today. It is apparent that they experienced quite rough handling during a trip estimated to cover 100 to 150 miles.

Toward the end of the Triassic period, a layer of plant debris about two hundred yards thick built up in this way. This layer is called the Chinle Formation. Then new geological events set in: The valley gradually dropped, and an even deeper layer of sediment about 1,000 yards deep accumulated over millions of years atop the original Chinle Formation. And while the land masses receded further, large areas of Arizona were finally flooded by shallow ocean waters. In this way, the tree trunks of the Triassic forests apparently found an eternal grave.

This development, sketched in a few words, took up not less than 100 million years of the earth's history. Then an entirely new event occurred: the birth of the Rocky Mountains. Violent forces lifted masses of land of today's Southwest, the sea receded, and the Chinle Formation, once under sea level, was again—after millions of years—at an altitude of 1,500 feet.

Then the tireless forces of erosion started their work. Wind and rain reduced the original sediment until the Chinle Formation was exposed once more. The trunks buried in it are too hard to be affected by erosion; but the sand and gravel around them washed away, bringing into view these gorgeous remains from the earth's past. The process is still going on. It will take a few more millions of years to erode the whole Chinle Formation. As a result, more and more agatized wood will be laid bare.

During petrification, minerals in solution in the surrounding water enter the wood in an infinitely slow process, filling the cells and intercellular spaces of the tissue. Silica (SiO_2) is the predominant material in the fossil trunks. The source of this material is found in the extensive deposits of volcanic ash in an area of the American West which, up to our own day, has experienced many violent volcanic shocks. Silica in itself is a colorless compound. The beautiful colors of the fossil wood are due to comtamination by other minerals during the petrification process. The splendid shadings of red, brown, and yellow indicate the presence of iron oxide; the deep brown and black colorings are caused by manganese oxide.

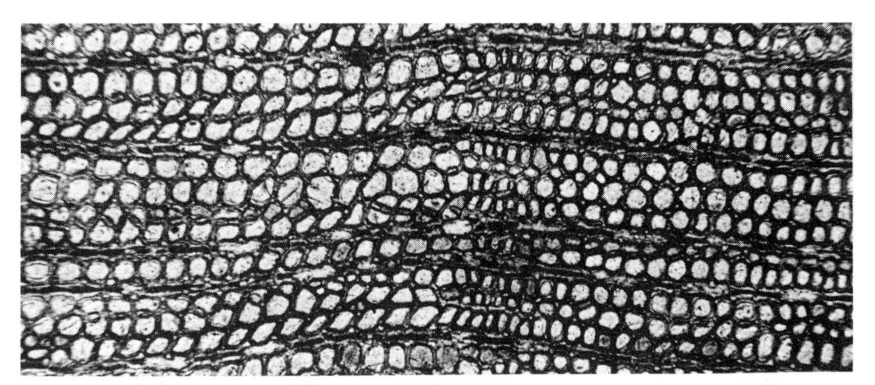

FIGURE 10-5. Cross section of petrified wood, 120×.

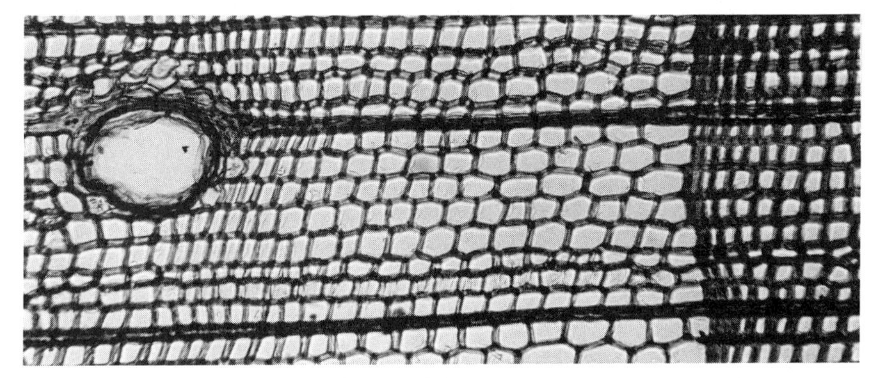

FIGURE 10-6. Cross section of a recent pine, 120×.

The microscopist, of course, is mainly interested in the question of how well the original anatomy of the wood has been preserved. The quality of petrification depends on the conditions under which the trees were inundated. Immediate exclusion of oxygen is necessary in order to prevent decay; then the structure of the wood can be preserved in the minutest detail.

Some trunks still show their growth rings. And the billions of small chips on which the visitors trample still look like wood. A cross section of a thumb-sized petrified splinter (see Figure 10-5), when compared with a cross section of *Pinus strobus*, the white pine, so common in today's forests (see Figure 10-6), confirms the perfect conservation of the cell structure. One difference is that the modern pine has resin ducts.

The most prevalent tree among the petrified blocks is the early pine, *Araucarioxylon arizonicum*, from which the cross section shown in Figure 10-5 was made. It has been extinct for a long time. Changes in climate have undermined the life-base of this tropical tree. Distant relatives are the *Araucarias* that live in South America, Australia, and on some of the South Sea Islands as *Araucaria excelsa*, the Norfolk Island pine.

Resources

Bringing this book to a close, with the hope that it will give the interested naturalist a head start in making the invisible visible, a few practical hints are in order.

One of the problems micronauts will face at the very beginning is how and where to get material for observation, study, experimentation, or photography, unless it is available in their immediate environment—the kitchen, the supermarket, or the pharmacy. In order to collect pond life, trips to the country are the most adventurous because what is caught can only be disclosed after coming home, unless a small field microscope is taken along. (Many field microscopes are optically quite inferior.) The thrill of collecting is compounded by the fact that the population of animalcules in ponds, ditches, streams, or lakes changes with the seasons. You never know what's in there. The surprise element is most rewarding. Collecting trips are of little use, however, if a particular species is wanted. A search from pond to ditch and ditch to pond will most likely end with failure.

Fortunately, the more common organisms, such as those frequently needed in biology classes, are available from biological supply houses. There are a number of companies that keep cultures of various organisms on hand for immediate delivery. Such companies can provide bacteria, protozoa, algae, insects, and nematodes, as well as media to culture these organisms at home or in the laboratory. A few of the microorganisms shown and discussed in the second part of this book were collected from nature, but more were bought from one of these companies, then subcultured in the author's laboratory. In this respect, the functions of these companies are of inestimable value. It is, for example, very unlikely that one will find the giant amoeba *(Pelomyxa carolinensis)* in nature. For all practical purposes, this amoeba, so important in research, is kept alive artificially in supply houses and laboratories.

The largest companies in the United States equipped to supply biological materials are the following:

Carolina Biological Supply Company
2700 York Avenue
Burlington, NC 27215

Powell Laboratories
Division of Carolina Biological Supply Company
Gladstone, OR 97027

General Biological Supply House, Inc.
8200 South Hoyne Avenue
Chicago, IL 61620

Ward's Natural History Establishment
P.O. Box 1712
Rochester, NY 14603, or
P.O. Box 1749
Monterey, CA 93940

The catalogues of these companies are heavy volumes offering, in addition to living material, preserved animals and plants, stained preparations (whole mounts or sections), objects of human, animal, and plant histology (normal and pathological), glassware, laboratory gadgets, chemicals, microscopes, and more.

Another consideration is personal contact with experienced microscopists, which is possible only where microscopical societies exist. In this respect the possibilities in the United States, even in densely populated areas, are very restricted as compared with the situation in some European countries, such a Germany and England, where a number of cities sustain such associations. In the United States, the following societies can be contacted:

The New York Microscopical Society has its headquarters at the American Museum of Natural History, 15 West 77th Street, New York, NY 10024. Members and guests meet once a month, September to June. During these evening gatherings, lectures on a variety of subjects are presented, experiences exchanged, problems discussed. The New York group also organizes workshops two or three times a year. These are held on Saturdays. Each course consists of eight to ten sessions. Each session provides theoretical instruction by specialists in the morning, followed by practical exercises in the afternoon. The New York Microscopical Society also issues a Newsletter two or three times a year. The annual membership fee is $20.00

The North-Eastern Association of Microscopy can be contacted c/o Ms. Barbara Forster, 53 Eton Street, Springfield, MA 01108. Its activities are also concerned with the arrangement of courses.

Both *The State Microscopical Society of Illinois,* 2001 N. Clark Street, Chicago, IL 60614, and *The Los Angeles Microscopical Society,* George C. Page Museum, 5801 Wilshire Boulevard, Los Angeles, CA 90036, also have as their purpose the teaching of microscopy.

The American Microscopical Society, a national, or even internationl organization, is strictly professional in character. It meets only once a year, between Christmas and New Year's Day, rotating its meeting places. Its members are almost exclusively biology professors at universities and colleges. The Society publishes the distinguished "Transactions of the American Microscopical Society," a quarterly periodical. Inquiries regarding membership may be addressed to the secretary. Since the office of secretary is held by different individuals during different years, the name and address of the present secretary can be determined by looking up a recent issue of the "Transactions," which can most likely be found on the library shelves of universities or colleges. The yearly membership fee is $24.00.

An important teaching institute is *The McCrone Research Institute,* 2820 S. Michigan Avenue, Chicago, IL 60616. This Institute regularly offers courses on various branches of light microscopy, such as photomicrography, polarized light microscopy, crystallography, and other specialties. Most of these are one-week courses, half devoted to theory, half to laboratory practice with the appropriate equipment. The McCrone Institute also publishes a journal, "The Microscope," under the imprint of "Microscope Publications, Ltd." It specializes in problems and techniques of industrial microscopy.

Finally, a word on *The Biological Photographic Association.* This is also an international association with many chapters throughout the United States. Although this is primarily an association of medical photographers, the majority of its members are also well trained in photomicrography. The BPA publishes a membership list of about 1200 members scattered throughout 35 chapters. This opens the prospect of establishing contact with experienced photomicrographers. The membership fee is $65.00 a year. More information can be obtained by writing to the executive secretary, Ms. Diane D. Jones, 1 Buttonwood Court, Indian Head Park, IL 60525. The BPA publishes a journal, "The Journal of Biophotography," four times a year. It is sent to members free of charge.

All of the societies and associations mentioned here offer membership at reduced fees for students.

References

REFERENCES FOR CHAPTER 1

Bradbury, S., *The Evolution of the Microscope*. New York: Pergamon Press, 1967.

Clay, R. S., and T. H. Court, *The History of the Microscope*. London: Charles Griffin & Comp. 1932.

Dobell, Clifford, *Antony van Leeuwenhoek and His "Little Animals."* London: Staples Press Ltd., 1932. (Also available as a paperback edition from Dover Publications Inc., New York.)

Hooke, Robert, *Micrographia*. London: Royal Society of London, 1665. (Also available as a paperback from Dover Publications, New York.)

REFERENCES FOR CHAPTER 2

Allen, Roy M., *The Microscope*. New York: Van Nostrand, 1940.

Barer, R., *Lecture Notes on the Use of the Microscope*. Springfield, Ill.: Charles C. Thomas, 1956.

Carpenter, William B., *The Microscope and its Revelations*. Philadelphia, Pa.: P. Blakiston's Son and Co., 1901.

Gage, Simon H., *The Microscope*. Ithaca, N.Y.: Comstock Publishing Co., 1947.

Hartley, W. G., *Hartley's Microscopy*. Charlbury, England: Senecio Publishing Co., Ltd., 1979.

Needham, George H., *The Practical Use of the Microscope, Including Photomicrography*. Springfield, Ill.: Charles C. Thomas, 1958.

Rochow, Theodore G. and Eugene G. Rochow, *An Introduction to Microscopy by Means of Light, Electrons, X-Rays, or Ultrasound*. New York: Plenum Press, 1978.

REFERENCES FOR CHAPTER 3

For Rheinberg Differential Color Illumination:

Delly, John G., *"Rheinberg Differential Color Photomicrography."* In: *Biomedical Photography, A Kodak Seminar in Print*, Eastman Kodak Publication N-19, Rochester, N.Y., Eastman Kodak Company, 1976, pp. 4–16.

Vetter, J. P., *The Production and Use of Rheinberg Differential Color Filters. J. Biol. Photographic Assoc.*, 31, 1963, pp. 15–18.

For Polarized Light:

Bennett, H. S., "The Microscopical Investigation of Biological Materials with Polarized Light." In: R. M. Jones, ed., *McClung's Handbook of Microscopical Techniques*, 3rd ed. New York, N.Y.: Hafner Press, 1950, pp. 591–677.

For Modulation Contrast:

Hoffman, Robert, "The Modulation Contrast Microscope. Principles and Performance." *J. Microscopy*, 110:(3), August, 1977, pp. 205–222.

For Phase Contrast:

Bennett, A. H., Helen Jupnik, Harold Osterberg, and Oscar W. Richards, *Phase Microscopy, Principles and Applications*. New York: John Wiley and Sons, 1951.

Gravé, Eric V., A Substitute Phase Contrast Attachment, Transactions of the American Microscopical Society, Vol. 96 (3), 1977, pp. 60–64.

For Nomarski Differential Interference Contrast:

Allen, R. D., G. B. David, and G. Nomarski, "The Zeiss-Nomarski Differential Interference Equipment for Transmitted-Light Microscopy." *Zeitschrift fuer wissenschaftliche Mikroskopie und mikroskopische Technik,* Vol. 69, No. 4, 1969, pp. 193–221 (in English).

For Interference Contrast after Jamin-Lebedeff:

Piller, H., *Interference Microscopy with Transmitted Light.* Carl Zeiss Publication G40-560/1-e, Oberkochen, West Germany: Carl Zeiss, 1963.

For Fluorescence Microscopy:

Holz, H. M., *Worthwhile Facts About Fluorescence Microscopy.* Carl Zeiss Publication K 41-005-e, Oberkochen, West Germany: Carl Zeiss, 1975.

REFERENCES FOR CHAPTER 4

Brown, G. D. and J. Creedy, *Experimental Biology Manual.* London: Heinemann Educational Books Ltd., 1970.

Corrington, Julian D., *Working With the Microscope.* New York: McGraw-Hill, 1941.

Guyer, Michael F., *Animal Micrology, Practical Exercises in Zoological Micro-Technique,* 5th ed. Chicago: The University of Chicago Press, 1953.

Metzner, Jerome and Philip Goldstein, *Experiments with Microscopic Animals.* New York: Doubleday, 1971.

Needham, James G., et al., *Culture Methods for Invertebrate Animals.* New York: Dover, 1959.

REFERENCES FOR CHAPTER 5

Allen, Roy M., *Photomicrography.* New York: Van Nostrand, 1941.

Eastman Kodak Company in cooperation with John G. Delly, *Photography Through the Microscope.* Rochester, N.Y.: Eastman Kodak Co., 1980.

Loveland, R. P., *Photomicrography,* Vols. 1 and 2, New York: John Wiley and Sons, 1970.

Needham, George H., *The Practical Use of the Microscope, Including Photomicrography*. Springfield, Ill.: Charles C. Thomas, 1958.

Shillaber, Charles P., *Photomicrography in Theory and Practice*. New York: John Wiley and Sons, 1944.

REFERENCES FOR CHAPTER 6

Calkins, Gary N., "*Didinium nasutum,* the Life History." *J. Exper. Zool.,* 19, 1915, pp. 225–239.

Giese, Arthur C., *Blepharisma*. Stanford, CA: Stanford University Press, 1973.

Kudo, R. R., *"Pelomyxa carolinensis Wilson,"* *J. Morphol.* 78, 1946, pp. 317–343.

Lloyd, Francis E., *The Carnivorous Plants*. New York: The Ronald Press, 1942.

REFERENCES FOR CHAPTER 7

Beers, C. D., "The Excystment Process in the Ciliate *Didinium nasutum.*" *J. Elisha Mitchell Sc. Soc.,* 61, 1945, pp. 264–275.

______ "The Excystment in *Didinium nasutum* with Special Reference to the Role of Bacteria." *J. Exper. Zool.,* 103, 1946, pp. 201–231.

Brønsted, H. V., *Planarian Regeneration*. New York: Pergamon Press, 1969.

Giese, Arthur C., *Blepharisma*. Stanford, CA: Stanford University Press, 1973.

Goss, Richard J., *Principles of Regeneration*. New York: Academic Press, 1969.

Hyman, L. H., *The Invertebrates,* Vol. 3. New York: McGraw-Hill, 1951.

Kidder, G. W. and C. L. Claff, "Cytological Investigations of *Colpoda cucullus.*" *Biol. Bull.,* 74, 1938, pp. 178–197.

Wichterman, Ralph, *The Biology of Paramecium*. New York: The Blakiston Co., Inc., 1953.

REFERENCES FOR CHAPTER 8

Ahmadjian, Vernon, *The Lichen Symbiosis*. Waltham, MA: Blaisdell Publishing Co., 1967.

Becker, Elery R. and Mary Talbott, "The Protozoan Fauna of the Rumen and Reticulum of American Cattle," *Iowa State College Journal of Science*, Vol. 1, October, 1926, pp. 345–354.

Buchner, P., *Endosymbiosis of Animals with Plant Microorganisms*. New York: Interscience Publishers, 1953.

Borel, Pierre, *Observationum Microscopicarum Centuria*, den Haag, 1656.

Buller, Reginald, *Researches on Fungi*, Vol. 6, London: Longmans Green and Co., 1934.

Cleveland, L. R., "The Wood-Feeding Roach *Cryptocercus punctulatus*, Its Protozoa, and the Symbiosis Between Protozoa and Roach." *Mem. Amer. Acad. Arts and Sciences*, 17(2), 1934, pp. 187–342.

Gravé, Eric V., "Observations on the Intestinal Flagellates of the Woodroach Cryptocercus punctulatus." *Microscopy* (London), 34, 1980, pp. 119–130.

Jennings, H. S., *Behavior of the Lower Organisms*. Bloomington, Ind.: Indiana University Press, 1962.

Keeble, Frederick, *Plant-Animals, A Study in Symbiosis*. Cambridge, England: Cambridge University Press, 1912.

Keeble, F. and Gamble, F. W., "The Origin and Nature of the Green Cells of *Convoluta roscoffensis*," *Quart. J. Microsc. Sc.*, Vol. 51, Pt 2, 1907, pp. 167–219.

Trager, William, *Symbiosis*. New York: Van Nostrand, 1970.

REFERENCES FOR CHAPTER 9

Creaser, Charles, "Structure and Growth of Scales of Fishes." Michigan University Museum of Zoology, Misc. Publ. no. 17, 1926, pp. 1–84.

Jennings, H. S., *Behavior of the Lower Organisms*. Bloomington, Ind.: Indiana University Press, 1962.

Tunison, A. V., et al., "Coloration of Fish." *Progressive Fish Culturist*, Vol. 9, No. 1, 1947, pp. 53–61.

REFERENCES FOR CHAPTER 10

Baross, John A. and Jody W. Deming, "Growth of 'Black Smoker' Bacteria at Temperatures of at Least 250°C." *Nature* 303, 6, 1983, pp. 423–426.

Chapman, Valentine J., *The Algae*. New York: St. Martin's Press, 1962.

Copeland, Joseph J., "Yellowstone Thermal Myxophyceae." *Annals of the New York Academy of Sciences*, Vol. 36, 1936, pp. 1–229.

Ransom, Jay Ellis, *Petrified Forest Trails, Guide to Petrified Forests of America*. Portland, Or.: Mineralogist Publishing Co., 1955.

BOOKS ON PROTOZOOLOGY

For Identification of Species and General Information

Corliss, John O., *The Ciliated Protozoa*. New York: Pergamon Press, 1979.

Curtis, Helena, *The Marvelous Animals, An Introduction to the Protozoa*. Garden City, N.Y.: The Natural History Press, 1968.

Jahn, Theodore L. et al., *How to Know the Protozoa*. Dubuque, Iowa: Wm. C. Brown Co., 1979.

Kirby, Harold, *Materials and Methods in the Study of Protozoa*. Berkeley and Los Angeles: University of California Press, 1950.

Kudo, Richard R., *Protozoology*, 4th ed. Springfield, Ill.: Charles C. Thomas, 1960.

Manwell, Reginald, *Introduction to Protozoology*. New York: St. Martin's Press, 1961.

Glossary

DEFINITION OF MICROSCOPICAL TERMS[1]

Abbe condenser: *see* Condenser, Abbe

Aberration, spherical: An imperfection in uncorrected lenses impairing the sharpness of the image. It is caused by the fact that light rays passing through a lens's periphery have a different focal point than those passing closer to the center. Central rays are not affected.

Aberration, chromatic: An imperfection similar to the spherical aberration, but caused by the fact that light rays of different wavelengths, that is, of different colors, come to a focus at different focal points.

Analyzer: A Polaroid filter used in combination with a polarizer to produce polarized light.

Angstrom: A unit of length equal to 10 millionths of a millimeter (symbol Å).

Aplanatic: A term applied to microscope condensers. Such condensers are adjusted to diminish spherical aberration.

Achromatic objectives: Objectives that are adjusted to eliminate, at least in part, spherical aberration in the 500 to 600 nm wavelength range of the spectrum (green-yellow) and chromatic aberration of the red and blue-violet, the two colors at both ends of the visible spectrum.

Apochromatic objectives: Objectives that are corrected for two colors spherically and three colors chromatically.

Chromatic aberration: *see* Aberration, chromatic.

Condenser, Abbe: A condenser invented by Ernst Abbe, a German physicist. It consists of two lens elements, the top of which is removable for low-power observation or photography.

Condenser, darkfield: A condenser constructed in such a way that by placing a stop either below or on top or inside the lens combination, direct light is prevented from illuminating the specimen. Depending on the stop and the objective used, only peripheral light can enter the objective. As a result, the object appears bright on a dark background.

Curvature of field: A characteristic of poorly corrected lenses to produce images that are curved, that is, in fair focus in the center, but losing sharpness progressively toward the periphery or vice versa. In plan-achromats and plan-apochromats, this aberration is corrected. Curvature of field is not to be confused with spherical aberration. The latter affects the image as a whole, not just part of it.

Depth of field: The image of an object can be sharp only within a certain range, depending on the optical system's depth of field. If the object's thickness exceeds this zone, it cannot be in overall good focus.

Diaphragm: An opening, fixed or variable, to control the amount of light to be admitted to an optical system. It is now used as an iris diaphragm consisting of adjustable metal blades. A good microscopical set-up has two diaphragms: the field diaphragm at the microscope lamp and the aperture diaphragm in the condenser. Some objectives have a built-in diaphragm inside the objective.

Eyepiece: This is the lens combination near the observer's eye. It is the component of a compound microscope that magnifies the primary image of the objective.

Glare: Excessive light that has a detrimental effect on the image formation.

Illumination, critical: A method of illuminating the specimen by focusing the image of the light source in the focal plane of the specimen. It is also named after the originator, Edward M. Nelson.

Illumination, Köhler: A method of illuminating the specimen by focusing the image of the light source on the aperture diaphragm of the condenser while focusing the field diaphragm on the plane of the object. It was originated by August Köhler.

Macrograph: The reproduction of a small object at a magnification of up to 10×.

Magnification: The product of the ×-value of the objective times the ×-value of the eyepiece. In photomicrography, the camera factor is to be considered. It reduces the magnification by ⅓ or ½, as the case may be. The photographic enlargement is to be taken into account for the final picture.

Mechanical Stage: An accessory to move the specimen in the north-south or east-west directions, useful if the field of view has to be changed by small distances that are difficult to achieve manually.

Micrograph: The reproduction of a small object beyond the range of a macrograph.

Micrometer: *see* Micron.

Micron: A unit of length equal to one thousandth of a millimeter (symbol μ). Also called micrometer (symbol μm).

Millimicron: *see* Nanometer.

Nanometer: A unit of length equal to one millionth of a millimeter (symbol nm), or one thousandth of a micron, also known as a millimicron (symbol mμ).

Nelson illumination: *see* Illumination, critical.

Numerical Aperture: The total of a formula expressed as a number, often engraved on the lens, that indicates the light-gathering and resolving power of an objective.

Ocular: *see* Eyepiece.

Pinhole eyepiece: An eyepiece deprived of all optical elements but capped by a disc with a small hole in the center, about one millimeter in diameter, for observing the backlens of an objective.

Polarized light: Light waves, vibrating in all directions can be made vibrating in one plane. Light waves are then "plane polarized." Originally achieved with special calcite prisms (Nicol prisms), it is now produced with two Polaroid filters, the analyzer and the polarizer.

Polarizer: A Polaroid filter used in combination with an analyzer to produce polarized light.

Resolution: The ability of an objective to form distinguishable images of structures separated by a small distance.

Spherical aberration: *see* Aberration, spherical.

Working distance: The distance between the coverglass or the uncovered specimen and the front lens of the objective.

DEFINITION OF BIOLOGICAL AND PHOTOGRAPHIC TERMS

Acetic acid: An acrid, irritating acid, harmful in concentrated form. A frequent ingredient of microbiological formulas.

Amino acids: The "building blocks" of the proteins. They are extremely complex organic compounds. A protein molecule consists of hundreds of amino acid molecules.

Anabolism: *see* Metabolism.

Anode: *see* Electrode.

Axostyle: An organ of filamentous, rod- or bandlike structure in certain flagellates. It serves as a support, or as a means of locomotion, or as an instrument to change a cell's shape.

Catabolism: *see* Metabolism.

Cathode: *see* Electrodes.

Chlorophyll: The green pigment in plant cells that enables the plant to synthesize carbohydrates from carbon dioxide and water under the influence of sunlight, a process known as photosynthesis.

Ciliates: A group of ciliated unicellular protozoa.

Clone: The descendants produced through cell division from a single animal. Members of a clone are of the same genetic constitution.

Commensal: An organism that lives with a dissimilar organism to its advantage without giving anything in return, but also without harming the host.

Compound eyes: The eyes of insects and crustaceans composed of a few to thousands of single units.

Conjugation: A sexual process among some protozoa in which the conjugants join in order to exchange nuclear material in a temporary union.

Contractile vacuole: An organ in protozoa that controls the water content of a cell. It expands if there is excess water in the cytoplasm, then contracts to expel water from the cell.

Crystals: The solid form of an element or a chemical compound growing undisturbed from a solution or from a melted condition. In solidifying, the atoms arrange themselves according to a definite structure that is reflected in smooth surfaces and geometrically correct angles.

Cytoplasm: The protoplasm of a cell excluding the nucleus or nuclei.

Cytostome: The mouth opening in protozoa.

De-differentiation: The process opposite to differentiation in which a cell's organelles lose their function, changing into an undifferentiated resting state.

Denizen: An inhabitant.

Detritus: Debris resulting from decomposing organic matter occurring in any water accumulation in nature, also regularly present in infusions.

Differentiation: The process of change in a developing organism in which its organs and shape are gradually formed.

Electrodes: The two terminals by which an electric current enters and leaves a liquid- or gas-filled vessel or vacuum tube. The positive pole is called the anode, the negative pole the cathode. The cathode emits so-called cathode rays, a stream of electrons, minute subatomic particles.

Emulsion: A liquid or substance in which small particles are evenly suspended. In the coating of photographic film or plates, photosensitive particles of silver bromide are embedded.

Enzymes: Chemical compounds that function as catalysts, for example, accelerating or promoting processes of metabolism, as digestion. Saliva, for instance, contains an enzyme to prepare and to speed up digestion. In catalyzing, an enzyme itself is not consumed, but can be reused in another reaction of the same kind.

Exoskeleton: The supporting and protecting shell of arthropods, such as insects, spiders, mites, crustaceans, and others.

Fermentation: The breakup of organic substances by microorganisms as bacteria, yeasts, or molds.

Food vacuoles: Fluid-filled cavities in a cell in which the digestion of ingested food takes place.

Foraminifera: A group of marine protozoa that build a shell of lime around themselves.

Genus (pl. genera): Classification name for animals and plants. Each genus consists of a number of similar species. Similar genera are grouped in a family.

Geotaxis: The response of an organism or a cell to the gravitational force of the earth.

Holophytic feeding: Feeding like a green plant through photosynthesis.

Holozoic feeding: Feeding like an animal by ingesting food.

Hyphae: The filamentous parts of a fungus that, if interwoven, form its mycelium.

Infrared: Rays of a wavelength 750 nm and more beyond the red end of the visible spectrum. They range from 750 nm to 0.03 cm.

Infusoria: Term for microorganisms, mainly small ciliates, found in infusions of organic substances, such as hay, soil, or manure.

Ingesta: Whatever material that is taken in as food.

Macronucleus: The organ in some classes of ciliates (as *Paramecium*) that functions as a control center for the chemical changes necessary to maintain the cell's life, essentially its metabolism. It is concerned with the vegetative tasks of life. (*See* Micronucleus.)

Medium: A nutritive solution or substance for cultivating microorganisms, such as bacteria, protozoa, algae, fungi, or tissues.

Metabolism: The chemical processes in a living organism related to the building up of the protoplasm and its conversion into energy needed by its activities. Constructive metabolism is called anabolism; destructive metabolism is catabolism.

Metamorphosis: The transformation of an animal during its development to the adult form. In many insects, for example, the changes are from egg to larva to pupa to adult.

Micronucleus: The organ in some classes of ciliates (as *Paramecium*) that plays an important role in the reproductive process of these protozoa. (*See* Macronucleus.)

Microphotograph: The photograph of a large object reduced, as a newspaper page on microfilm.

Mitochondria: Minute spherical or rod-shaped structures present in all cells (except bacteria and blue-green algae). They contain many enzymes and are involved in the cells' metabolism.

Mycelium: *see* Hyphae.

Myoneme: A filament capable of expansion and contraction present in many protozoa as in the ciliate *Lacrymaria olor*. A primitive muscle.

Organelles: Structures in unicellular animals and plants that perform specific functions, as the flagella, or cilia, which serve for locomotion, or the contractile vacuoles, which discharge excess water from the cell.

Osmosis: A process based on the semi-permeability of cell membranes. It permits the passage of fluids through their skin. Protozoa that cannot ingest solid food depend on this process for their nourishment.

Parasite: An organism that lives with another dissimilar organism from which it receives food or shelter without giving anything in return, even harming, sometimes killing the host.

Peristome: A twisted opening leading to the mouth of various protozoa as, for example, *Bursaria truncatella*.

Photomacrograph: The photograph of an object at a magnification of up to $10\times$.

Photomicrograph: The photograph of a small object enlarged beyond the magnification of a photomacrograph.

Photosynthesis: A process in green plants in which carbohydrates are synthesized from carbon dioxide and water under the influence of sunlight.

Phototaxis: *see* Phototropism.

Phototropism: The sensitivity of an organism to light, also called phototaxis. The response can be positive, toward the light, or negative, away from it.

Proctolactic feeding: Feeding on excrements.

Protein: Very complex organic compounds present in all living matter. *See also* Amino acids.

Protista: Term for all unicellular organisms, animals, plants, and bacteria.

Protoplasm: The substance within a cell including the nuclei and the cell membrane but excluding ingested matter.

Pseudopod: A temporary protrusion of a cell, as of an amoeba, that serves as a means of locomotion and feeding. It can also be withdrawn at will.

Reciprocity failure: States the fact that the reciprocity law is not valid for very long or very short exposures. It can be corrected with compensating color (CC) filters. The effect is less pronounced in black-and-white than in color photography.

Reciprocity law: The photochemical law stating that the effect of light on a photosensitive emulsion is the product of the light intensity multiplied by the exposure time. Both values are reciprocal. An increase in light intensity requires a reduction in exposure in order to get the same effect on the emulsion, and vice versa.

Rhombohedron: A six-sided prism having opposing parallel faces. A cube is a special case having face angles of 90°, neither acute nor obtuse.

Rotifer: A class of microscopic aquatic animals. They are metazoa consisting of many cells. Yet in size they are not larger than *Paramecium*. They are provided with a ciliated circle at the "head." The cilia move one after the other, giving the impression of a rotating wheel. The cilia serve the animal for locomotion and feeding.

Slipperanimalcule: Popular name for *Paramecium* derived from this animal's resemblance to a slipper.

Soredia (plural of Soredium): Organs of reproduction in lichens. Lichens, consisting of two plants, an alga and a fungus, form special units in which a few algal cells are entangled within the hyphae of the fungus. Freed from the parent plant and spread by the wind, they may become a new plant if they find favorable conditions for growth.

Stomata: Minute pores in the epidermis of plants, particularly in leaves, through which gaseous exchanges take place.

Substrate: The base on which a plant grows or an animal walks or to which it is attached.

Symbiont: A partner in a symbiotic relationship.

Symbiosis: The association of two dissimilar organisms to the benefit of both partners.

Tannic acid: A chemical with a strongly astringent property.

Trichites: Needlelike structures occurring in ciliates around their mouth openings. They serve as a reinforcing device.

Trichocysts: Minute needlelike structures embedded in the pellicle of many ciliates. Their function is not known, but believed to be defensive because they are often released to form a screen around the attacked protozoan. They are then shot out and appear as slender rodlets surrounding the animalcule.

Vascular bundle: The part of tissue in higher plants through which water, salts, and other nutrients are transported throughout the plant. It also serves as a support. Wooded parts of a tree consist mainly of vascular bundles.

Wheelanimalcule: *see Rotifer.*

Notes

CHAPTER 1

1. S. Bradbury, *The Evolution of the Microscope,* (Oxford: pub, 1967), p. 3.

2. William B. Carpenter, *The Microscope and Its Revelations,* 8th edition, (Philadelphia: Blakiston's Son & Co., 1901), p. 126.

3. Reginald Clay and Thomas Court, *The History of the Microscope,* (London: pub, 1932) p. 11.

4. Clifford Dobell, *Antony van Leeuwenhoek and his "Little Animals",* (London: pub, 1932), p. 37.

5. ibid. pp. 40–41.

CHAPTER 2

1. Very recently, in 1982, Nikon developed high-power objectives that can be used dry. One provides a 60×, another one a 100× magnification. They require, however, Nikon eyepieces.

CHAPTER 3

1. Deborah Scarff, "Economical Procedure To Convert a Standard Light Microscope to the Hoffman Modulation Contrast System," *Journal of Biological Photography,* 50, no. 4, (October 1982), pp. 131–34. Peter Hoffmann, "Modulationkontrast-Verfahren im Eigenbau," *Mikrokosmos,* 72, no. 2, (February 1983), pp. 58–62.

2. Dieter Gerlach, *Das Lichtmikroskop*, (Stuttgart: Georg Thieme Verlag, 1976), pp. 217–218.

CHAPTER 4

1. James G. Needham, et al., *Culture Methods for Invertebrate Animals*, (New York: Dover, 1959), pp. 1–590. A valuable compendium that contains several hundred formulas for different invertebrates, their collecting, culturing, maintaining, and so on.

2. Edgar P. Jones, "Paramecium multimicronucleatum: Massculturing, Maintenance and Rehabilitating etc.", in *Culture Methods for Invertebrate Animals*, ed. James G. Needham, (New York: Dover, 1959), pp. 122–127.

3. Catalogue number L50P, available in quantities of 12 pellets ($2.80) to 100 pellets ($11.14).

CHAPTER 5

1. The symbol ASA (American Standards Association) as an indication of the speed of a film emulsion, has been changed to ISO (International Standards Organization).

2. Frederick Wratten (1840–1926) was a manufacturer of photographic products and one of the first to produce photographic filters. His system of numbering is, with some exceptions, still retained. After his death, the Eastman Kodak Company took over his business and continues to manufacture the filters.

CHAPTER 6

Re: The Venus's Fly-trap

1. C. Darwin, *Insectivorous Plants*, (London: L. Murray, 1875), p. 462.

2. Frank Morton Jones, "The Most Wonderful Plant in the World," *Natural History*, 23 (1923), pp: 589–596.

3. John Ellis, "Description of a New Sensitive Plant, Called *Dionea muscipula* or Venus's Fly-trap," London, (1770), Lloyd, Francis Emmet, *The Carnivorous Plants*, New York: The Ronald Press, 1942, p. 210.

4. Francis Emmet Lloyd, *The Carnivorous Plants*, (New York: The Ronald Press, 1942), p. 352.

5. J. Burdon-Sanderson, "Note on the electric phenomena which accompany stimulation of the leaf of *Dionaea muscipula*, *Proceedings of the Royal Society of London*, 21:495–496, 1873. J. Burdon-Sanderson, "On the Electromotive Properties of the Leaf of *Dionaea muscipula* in the Excited and Unexcited States," *Philosophical Transactions of the Royal Society of London*, 173 (1882), pp. 1–53, and 179 (1888), pp. 417–449.

CHAPTER 7

1. Chen, T. T., "Conjugation in *Paramecium Bursaria*," *J. Morphology*, Vol. 79, 1946, pp. 125–262.

2. Ulysses, so goes the story in Homer's epic *The Odyssey*, became a prisoner of Polyphem, one of the Cyclopses, a race of one-eyed giants. After the giant had devoured six of Ulysses's twelve comrades, he got the giant drunk, then blinded him with a burning pole and escaped.

CHAPTER 8

1. Vernon Ahmadjian, *The Lichen Symbiosis*, Waltham, MA: Blaisdell Publishing Co., 1967, p. 78.

2. The British zoologists, F. Keeble and F. W. Gamble, who worked out the worm's life history at the turn of the century, estimated that two square yards of such a green patch may contain 5,600,000 individuals.

3. Cleveland, L. R., D. R. Hall, E. P. Sanders, and J. Collier, The woodfeeding roach, *Cryptocercus*, its protozoa, and the symbiosis between protozoa and roach, Memoirs of the American Academy of Arts and Sciences, Vol. 17, 1934, pp. 187–342.

4. Yamin, Michael A., Flagellates reported from Lower Termites and the woodfeeding roach *Cryptocercus*, Sociobiology, Vol. 4 No. 1, Chicago: California State University, 1979.

5. For a complete record of so far undescribed protozoa in the roach's hindgut, see my article, "Observations on the Intestinal Flagellates of the Woodroach *Cryptocercus punctulatus*," *Microscopy* (London) 34 (December 1980), pp. 119–130.

6. Translation from the Latin by Dr. Adolf Lamm.

CHAPTER 9

1. Philip Henry Gosse, *Evenings at the Microscope*, (London: Society for Promoting Christian Knowledge, 1859), p. 24.

CHAPTER 10

1. A. E. Walsby, "Bacteria that Grow at 250°C," *Nature*, Vol. 303, (1983), p. 381. Walsby is the discoverer of the square bacterium.

GLOSSARY

1. Courtesy American Society for Testing and Materials, 1916 Race Street, Philadelphia, PA 19103.

Index

Abbe, Ernst, 39
Abbe condenser, light paths in, 36
Acorns, transport by squirrel, 159
Agar, nutrient, 63
  collection of microorganisms on, 63
Ahmadjian, Vernon, 138
Air, microorganisms in, 62–63
Algae:
  alga, functions of, 137–38
  and *Chlorella,* 139
    crystals in, 139, 140
    culture experiment with, 139
    in darkness, 139–40
    and *Hydra viridis,* 140
    in *Paramecium bursaria,* 139
  fungus, functions, 137–38
  hyphae, 137, 138
  lichens, nature of, 139
  parasitism in, 138
  Parmela rudecta, 137
  reproduction, 138
  in sloth hairs, 138–39
    crevices, 139
Amati, Salvino degli, 4
American Microscopical Society, 178
American Museum of Natural History, 177
Amoeba. *See* Chaos chaos
Ångstrom, Anders J., 9
Araucarioxylon arizonicum, 175
Aresco, Inc., 42

Bacilli, mounting of, 62
Bacilli subtilis, on teeth, 62
Bacon, Roger, 3–4
Bacteria, temperature for growth, 63
Baross, John, 171
Beccaria, Pater, 113
Bedbug, proboscis of, 136
Bees, stings of, 134–35
Beginner, equipment for, 31–32
Belden Communications, Inc., 43
Biological Photographic Association, 178
Blepharisma americanum, 93–94
  cannibalism of, 93–94
  feeding, 93, 94
  food for, 93
  giantism of, 93
  and *Tetrahymena,* 93
Borel, Pierre, 4, 130, 156
Bradypus tridactylus, 138, 139
Brightfield illumination, 33–34
  picture using, 34
Buffon, 138
Buller, Reginald, 155
Burdock, burrs of, 159
Burdon-Sanderson, L., 110
Bursaria truncatella, 90–92
  cytostome, 90–91
  escape of prey, 91, 92
  feeding, 90
  ingestion of *Paramecia,* 91–92
  nature, 90

Carolina Biological Supply Company, 65, 177
Carpenter, William B., 5
Celandine seeds, elaiosomes of, 159
  and ants, 159

Centuria Observationem Microscopicarum, 4, 130, 156
Chaos chaos catching *Paramecium:*
advantages of laboratory use, 88
compared to leukocytes, 90
division, 87
vs. *Paramecia,* 88–90
pseudopods, 88, 90
size, 87
stimulation of, 87
streaming of, 87–88
vacuoles, 90
Cheek, cells from, 62
Cheese mite (*Tiroglyphus siro* L.), 156–58
Hypopus stage, 157
larva, legs of, 157
migration by, 157
Cleveland, L. R., 146
Cock, Christopher, 5
Cohn, Ferdinand, 103
Condensation, 57
Conjugation of *Paramecia:*
aurelia species, 120
bursaria, structure, 119
and *chlorella,* 120, 121
cilia, 122
clone, nature of, 120
clumping, 121–22
macronucleus, 119, 120
mating types, 120
micronucleus, 119, 120
nuclei, fusion of, 119–20
synkaryon, 120
Convoluta roscoffensis and *Chlamydomonas* spp.:
and algae cells, functions of in worm, 142
degeneration of, 143
reproductive cycle, 143
Chlamydomonas types, 142–143
digestion of, 141
on algae, 142
egg capsules, 142
eyespots, 140
flagellated algae, 143
green cells, 141
infection of worm, 143
nitrogen products, 141
otocyst, 140, 141
reproduction of, 141
Crabs, eyes of, 60–61
Crystals, 55–57
melt, growing from, 57
metaphen, 56
polarization, 55
salt, 56
shapes, 56
sugar, 55
tears, 56
in wax, 57
Crustaceans, 60
Cyclops, production of, 123–25
delivery of young, 125
egg sacs, 124
eggs, hatching time of, 123
female, 123
larvae, molts of, 125
Nauplius, larval stage of *Cyclops,* 124–25
Cysts, formation of and escape from for survival:
cilia, 116
in *Didinium,* 116
differentiation of structures, 118–19
durability, 119
endocyst, 116
enzyme for escape, 116
escape from, steps, 116, 117, 118
pictures of, 117–18
reasons for, 114
rotation of embryo, 116
time for, 119

Dandelion seeds, 158
Darkfield illumination:
coins as discs for, 37–38
Diploneis crabo, photographed with, 38, 39
disc for blocking light, 33, 36
filters, 37
Paramecium photographed with, 35, 39
powers, magnification, 39
procedure, 33, 36
Radiolaria photographed with, 38, 39
sizes, 37
stop, support for, 36–37
thickness of discs, 37
Darwin, Charles, 105, 106, 109
Diaphragm, 23, 26, 28
De Graff, Reinier, 6
Deming, Jody, 171
Descartes, quoted, 85
Didinium nasutum, 94–98, 114, 116–19
food search, 95
vs. *Paramecium,* 95, 96, 97
seizing organ of, 95
survival of, 114, 116–19
Differential interference contrast, Nomarskian, photography, 36
Division of animalcules:
Blepharisma americanum, 111, 112
nucleus, 112
organelles, stages of, 112–13
Dobbs, Arthur, 109
Dobell, Clifford, 8
Du Pont, Giovanni, 4

Ehrenberg, Christian Gottfried, 41
Electron microscopes, nature, 10
Ellis, John, 109
Evenings at the Microscope, 165
Eyeglasses, early, 4
Eyepieces, 23–24
elements of, 23
magnification, 23
vs. objectives, 23

Filters. *See* Photomicrography
Fish scales:

Fish scales *(cont.)*
aquarium, 167
changes in, reasons for, 166–67
chromatophores as half-tone screen, 166
flounder, 163–64
guanin crystals, 165
life growth, 167
melanophores, 164
Nile catfish, 166
patterns, 165–66
salmon, 165
salmon, Atlantic, example, 167–68
sardine, 164
seasons, 167
in skin, 163
source, 163
spawning mark, 168
stimuli to pigments, 166
xanthophores, 164
Fluorescence:
Acridine Orange, 51, 53
Auramine O, 53
barrier filter against ultraviolet, 52
BG3, 52
and black light, 51
exciter filter, 52
light source, 52
Neutral Red, 51
Nos. 50 and 52 glass, 53
Phloxinrhodamin, 53
primary, 51
Rhodamine B, 53
secondary, 51
stains for, 51
Forster, Barbara, 177
Freshwater specimens, perpetuation of, 64–65
Fungi, 63. *See also* Algae; *Pilobolus crystalinus*

Galileo, Galilei, 4
General Biological Supply House, 177
Gentian Violet, 62
Gerlach, Dieter, 52
Giese, Arthur C., 93
Gosse, Philip Henry, 165

Hairs, 61–62
Hay, 62
Herapath, W. B., 43
Hoffman, Dr. Robert, 47
Holz, H. M., 53
Hooke, Roger, 4–5, 15
invention of microscopes, 5
lenses used, 5
life, 5
Horse manure, 64
Hot Spring algae, in Yellowstone Park:
and Black Smokers, ocean floor, 171
blue-green, 169–70
Denticula thermalis, 172
evolution of, 170–71
mitochondria, 170
Phormidium alga, 171–72
plasm theory, 170
specimens, keeping of, 172
Synechococcus algae, 171
temperature tolerance, 170
thermophiles, proteins and amino acids of, 171
Hydra, catching waterflea, 100–01
and *Daphnia*, 100
nettling capsules, 100

Ibn Alhasan, Abu Ali al-Hazan, 3
Illumination:
Jamin-Lebedeff, 9
Nomarski, 9
Incident illumination:
combined with weak transmitted light, 47
illumination, 46
nature, 46
vs. reflected light, 46
reflections, control of, 46, 47
Indian cress seed, 158
Interference microscopes, 9, 50
with birefringent calcite plate, 50
colors, 50
differential, 50
Insects, 59–60
aphids, 60
compound eyes of, 60
fly, 60
killing jars, 59, 60
potashing, 60
Insects, hindguts of, parasites in:
axostyles, 149
Barbulanymoha ufalula, 147
cellulose, transformation to glucose, 144
Cryptocercus punctulatus, 145–46
functions, 145, 146
cysts, cycle, 146
flagellates, species of, 147
flora, 144
molting, 144
Pyrsonympha major, 145
Pyrsonympha vertens, 145
Reticulotermes flavipes, 144
Saccinobaculus ambloaxostylus, 147
specimens, preparation of, 149–50
Spirochaetes, 145
Trichonympha gracilis, 145
unknown forms, 147–49
Urinympha talea, 147
wood-feeding termites, 144
woodroach, 145, 146
intestinal tract of, 146

Janssen, Hans, 4
Jennings, H. S., 161
Johns Hopkins University, 171
Jones, Diane D., 178
Jones, Edgar P., 65
Jones, Frank Morton, 109

Keeble, F., 142
Kelvin, William T., 73

Kodak, Eastman, Company, 70
"Kodak Filters," 70
Köhler illumination:
  adjustment for, 26
  diagram, 28
Kudo, R., 65

Lacrymaria olor, 98–100
  finding, 98, 100
  pictures of, 99
  search of, for food, 98
  size, 98
Land, Edwin, 43
Leaf, epidermis of, 58–59
Leeuwenhoek, Antony van, 6–8, 113, 124
  discoveries of, 6, 8
  life of, 6
  light used by, 8
  microscopes of, 7–8
  publication by, 6, 7
Leidy, Joseph, 144
Lenses. *See* Optics
Lichen. *See* Algae
Light:
  nature, 9
  sources, for microscopes, 23–24
Linnaeus, Carolus, 87
Lobsters, 60
Los Angeles Microscopical Society, 178
Louis XIV, 130

Magnification:
  ancient work, 3
  dark ages, work on, 3
  limit, 22, 23
  medieval, 3–4
  range, 22, 23
Mast, S. O., 97
Maupas, E. F., 120
McCrone Research Institute, 178
Media, artificial, for growth:
  litmus paper for, 65–66
  pH adjustment, 66
  pH paper, measurement with, 65
  recipe, 65
  wheat medium, 66
Micrographia, 4, 16
"Microscope, The," 178
Microscope and Its Revelations, The, 5
Microscope, compound, invention of, 4
Microscope, parts and types of:
  automatic, 11
  basic parts, 12
  condenser, 13
  knobs, 12
  lenses, types of, 14
  light in, 11
  mirror, 13–14
  stage for slides, 13
  tube length, 13
Microscope, setting up of:
  condenser vs. focus, 26
  image, illumination of, 27
  iris diaphragm, 26, 28
  lamp, placement of, 24
  pinholes, 24, 26
Mirrors, 13–14, 40–41
Modulation contrast, 35, 47–48
  diagram, 47
  nature, 47–48
  photography with, 35
Modulation Optics, Inc., 47
Mosquito, proboscis of, 135–36
  saliva of, 135
Mounts, temporary, materials for, 54–55
Myonemes, 97

Nature, 171
Nature Photography: A Guide to Better Outdoor Pictures, 68
Nelsonian/critical illumination, 28–31
  diagram, 30
  example, 29
  focusing, steps, 30
  nature, 28
  procedure, 29, 31
Nettling capsules of *Hydra,* 132–33
New York Microscopical Society, 177
Nicol, William, 43
North-East Association of Microscopy, 177
Numerical aperture:
  diagram, 21
  effects, 21, 22
  formula, 20
  nature, 20

Oblique illumination:
  decentralized mirror, 40, 41
  discs, substage, 39, 41
  hole diaphragm, 40
  makeshift, 41
  triangle cutout, 40, 41
Ocean, specimens from, 64
Oldenburg, Henry, 6
Onion:
  bulb, 58
  calcium oxalate, 57
  skin, 57–58, 59
Optics, discussion and summary:
  achromats, 17
  apochromats, 17
  biconvex lens, 15
  chromatic abberation, 16, 17
  concave lens, 16
  cover glass correction oculars, 19–20
  diverging lens, 15
  field flatness, 19
  focal points, motion of with movement of light source and lens, 15
  glass to air, passage of light, 18, 19
  immersion, optical effects, 18, 19
  immersion objectives, reasons for, 18, 19
  lens types, 14
  objectives, data on, 20
  plan-achromats, 19
  refractive index, nature of, 14–15
  specimen, mounting of, 18, 19

Optics *(cont.)*
spherical aberration, 16, 17
Opus Majus, 3
Osolinski, Stan, 68
Oregon State University, 171

Page, George C., Museum, 178
Pallas, P. S., 127
Paramecium:
and algae, 139
vs. *Bursaria,* 91–92
vs. *Chaos chaos,* 88–90
conjugation of, 119–21
darkfield photography, 35, 39
vs. *Didinium,* 95–97
polarization, 45
Peters, B. G., 131
Petrified Forest, Arizona, 172–75
Chinle Formation, 174
compared to recent pine, 175
cross-sections, 175
discussion, 172–74
formation of, 174
history, 172–74
silica in, 174
Triassic period, 174
Phase contrast:
annular ring diaphragm, below condenser, 49
microscopes, 8–9
nature, 48
photography with, 35
ring, at back focal plane of objective, 49
systems for, 49
"Philosophical Transactions," 7
Photography Through the Microscope, 74
Photomicrography:
adapter, 69
author, setup of, 70
automatic exposure device, 69–70
black-and-white films, 76
CC filters, 76
color groups, 72
color photography, 73
cross hairs, 69
discovery, nature of, 81
discussion, 68
dust, problems with, 81
exposure:
light intensity, 78
meters, problems with, 78
specimen density, 78
film for, 68–69
film speed vs. conversion filters, 75
filters:
definition, 73
image contrast, 73
infrared, 72
light intensity, 72
performance of, 70, 72
fingerprints, 81
instant, 77
Kodachrome films, 74
Kodak filters, color film, 74
light sources, color temperatures of, 73
numbers on adapters, 69
out of focus, 80–81
Polaroid, 77
reciprocity failure, 76
roll length, 77
setups, trial and error, 78–79
sharpness check, 80, 81
spectrophotometric curves, 71
3200°K, 75
transparencies, 77
tungsten, 73, 74
uneven illumination, 80
Pilobolus crystalinus, fungus species:
experiments with, 155
firing angle, 154
glue on spore cases, 154
growth of, 155
in horse manure, 152, 153
light, aiming by, 154
power for, 154–55
pressure in, 155
sporangium, 156
spore cases, ejection of by, 153
structure, 153
swelling body, lens in, 155
views of, 153
Planaria flatworm, regeneration of:
fragments, 127, 128
neoblasts in, 128–29
organs of, 126
regeneration in evolutionary scale, 126
steps, 128–29
two-headed, 127, 128
wound, healing of, 127
Polarization:
anisotropic materials, 44
compensators, 44–45
crystals, 55
development, 43
granite slice, 45
interference colors, 44
isotropic materials, 44
nature, 43, 44
Paramecium, 45
photography with, 35
rotation of filters, 44
vitamins, 45
Pollen, pine, 158
Powell Laboratories, 177
Processionary caterpillars, poisonous hairs of, 133
Protozoa, activity of:
air bubbles, control of with, 66
detritus, control of with, 66
methanol, 67
narcotizing chemicals, 67
Protoslo, 67
Protozoology, 65

Quadruple division, *Colpodidae,* 113–15
Colpoda aspera, 113, 114

Quadruple division *(cont.)*
cycle, 113–14
cysts, 114
pictures, 115

Radiolaria, migration of, 159
Resolution:
limit, 9
vs. numerical aperture, formula, 22
Rheinberg, Julius, 42
Rheinberg illumination:
cellophane, 42, 43
colored rings on specimen, 42
Diazochrome, Tecnifax, 42
Kodak Wratten gelatine fibers, 42
Plexiglass, 43
Rotifer, 161
Royal Society of London, 6
Rumens of ruminants, protozoa in:
bacteria as symbionts, 150
cellulose, transformation of, 150
Diplodinium bursa, 151
Diplodinium dentatum, 151
infection by, 152
Isotricha prostoma, 151
life spans, 150–51
myonemes, 152
Ophryoscolecidae, 151–52
reproduction, 151–52
species, 151

Saturnulus elipticus, 159
Saussure, Horace, 113
Seneca, 3
Shock and trauma, protozoan reactions to:
discussion, 160
escape from, 162
habituation to, 161, 162
learning in, 163
Rotifer, 161
Stentor coeruleus, 160–61, 162
Stentor roeseli, 162
types of, 160
Slides, 13
Spallanzani, Lazzaro, 113
Spencer microscope lamp, 24, 25
and filament, image of, 25
Stains, 41–42, 55. *See also* Fluorescence
Starch grains, 57
State Microscopical Society of Illinois, 178
Stinging nettle, structure, 135
"Substitute Phase Contrast Attachment, A," 49
Supply houses, list, 177

Tannic acid, 97
Taxonomy, summary, 85–87
Tecnifax corporation, 42
Teeth, bacteria from, 62
Textiles, 59
Tobacco smoke, 57
"Transactions of the Microscopical Society," 49, 178
Trembley, Abraham, 126
Trichites, 97
Trichocysts, 97

Ultraviolet light, 9. *See also* Fluorescence
Utricularia, underwater animalcule capture by:
bristles of, action, 102, 103
digestion, 105–06
group of, 102
quadrifids in, 103, 104
trap of, functions, 102–03
setting of, 103
triggering of, 104
velum, 103
waterfleas, 104–05

Venus's flytrap, 106–10
closing, stages of, 108–09
electrophysiology of, 110
epidermis of, 109, 110
insect, catching of, 106, 107
leaves, 107
location of species, 106
sensitivity of trap, 107
size of prey, filtration by leaves of, 108, 109
structure, 106, 107
trigger hairs in, 110
Vinegar eel *(Anguillula aceti/Turbatrix aceti),* 129–32
acetic acid, 131
culture of, 131
filtration of, 131
pictures, 130
and vinegar, 130, 131
young of, 129
Viogen, 62

Walsby, A. E., 171
Ward's Natural History Establishment, 177
Wasps, poisons of:
Cerceris tuberculata Klg. vs. *Cleonus opthalmicus,* 134
Eumenes pomiformus Fbr. vs. caterpillars, 134
sting of, 134
Water, specimens from, 64
Water drops, as lenses, 61
Waterfleas. *See Utricularia*
Working distance, 21
Wratten filters, 42

Xanthophores, 164

Yamin, 146

Zeiss, Carl, 53
Zeiss Company, 11, 53
Zernicke, Frits, 8, 48